Science 500
Teacher's Guide

CONTENTS

Author: **Barry Burrus, M.S.**
Editor: Alan Christopherson, M.S.

Alpha Omega Publications ®
300 North McKemy Avenue, Chandler, Arizona 85226-2618

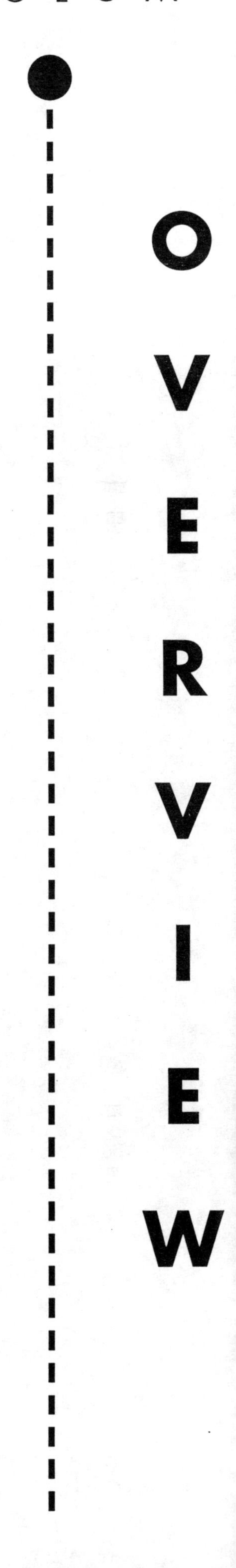
OVERVIEW

SCIENCE

Curriculum Overview
Grades 1–12

	Grade 1	Grade 2	Grade 3
LIFEPAC 1	YOU LEARN WITH YOUR EYES • Name and group some colors • Name and group some shapes • Name and group some sizes • Help from what you see	THE LIVING AND NONLIVING • What God created • Rock and seed experiment • God-made objects • Man-made objects	YOU GROW AND CHANGE • Air we breathe • Food for the body • Exercise and rest • You are different
LIFEPAC 2	YOU LEARN WITH YOUR EARS • Sounds of nature and people • How sound moves • Sound with your voice • You make music	PLANTS • How are plants alike • Habitats of plants • Growth of plants • What plants need	PLANTS • Plant parts • Plant growth • Seeds and bulbs • Stems and roots
LIFEPAC 3	MORE ABOUT YOUR SENSES • Sense of smell • Sense of taste • Sense of touch • Learning with my senses	ANIMALS • How are animals alike • How are animals different • What animals need • Noah and the ark	ANIMAL GROWTH AND CHANGE • The environment changes • Animals are different • How animals grow • How animals change
LIFEPAC 4	ANIMALS • What animals eat • Animals for food • Animals for work • Pets to care for	YOU • How are people alike • How are you different • Your family • Your health	YOU ARE WHAT YOU EAT • Food helps your body • Junk foods • Food groups • Good health habits
LIFEPAC 5	PLANTS • Big and small plants • Special plants • Plants for food • House plants	PET AND PLANT CARE • Learning about pets • Caring for pets • Learning about plants • Caring for plants	PROPERTIES OF MATTER • Robert Boyle • States of matter • Physical changes • Chemical changes
LIFEPAC 6	GROWING UP HEALTHY • How plants and animals grow • How your body grows • Eating and sleeping • Exercising	YOUR FIVE SENSES • Your eye • You can smell and hear • Your taste • You can feel	SOUNDS AND YOU • Making sounds • Different sounds • How sounds move • How sounds are heard
LIFEPAC 7	GOD'S BEAUTIFUL WORLD • Types of land • Water places • The weather • Seasons	PHYSICAL PROPERTIES • Colors • Shapes • Sizes • How things feel	TIMES AND SEASONS • The earth rotates • The earth revolves • Time changes • Seasons change
LIFEPAC 8	ALL ABOUT ENERGY • God gives energy • We use energy • Ways to make energy • Ways to save energy	OUR NEIGHBORHOOD • Things not living • Things living • Harm to our world • Caring for our world	ROCKS AND THEIR CHANGES • Forming rocks • Changing rocks • Rocks for buildings • Rock collecting
LIFEPAC 9	MACHINES AROUND YOU • Simple levers • Simple wheels • Inclined planes • Using machines	CHANGES IN OUR WORLD • Seasons • Change in plants • God's love never changes • God's Word never changes	HEAT ENERGY • Sources of heat • Heat energy • Moving heat • Benefits and problems of heat
LIFEPAC 10	WONDERFUL WORLD OF SCIENCE • Using your senses • Using your mind • You love yourself • You love the world	LOOKING AT OUR WORLD • Living things • Nonliving things • Caring for our world • Caring for ourselves	PHYSICAL CHANGES • Change in man • Change in plants • Matter and time • Sound and energy

	Grade 4	Grade 5	Grade 6
LIFEPAC 1	PLANTS • Plants and living things • Using plants • Parts of plants • The function of plants	CELLS • Cell composition • Plant and animal cells • Life of cells • Growth of cells	PLANT SYSTEMS • Parts of a plant • Systems of photosynthesis • Transport systems • Regulatory systems
LIFEPAC 2	ANIMALS • Animal structures • Animal behavior • Animal instincts • Man protects animals	PLANTS: LIFE CYCLES • Seed producing plants • Spore producing plants • One-celled plants • Classifying plants	ANIMAL SYSTEMS • Digestive system • Excretory system • Skeletal system • Diseases
LIFEPAC 3	MAN'S ENVIRONMENT • Resources • Balance in nature • Communities • Conservation and preservation	ANIMALS: LIFE CYCLES • Invertebrates • Vertebrates • Classifying animals • Relating function and structure	PLANT AND ANIMAL BEHAVIOR • Animal behavior • Plant behavior • Plant-animal interaction • Balance in nature
LIFEPAC 4	MACHINES • Work and energy • Simple machines • Simple machines together • Complex machines	BALANCE IN NATURE • Needs of life • Dependence on others • Prairie life • Stewardship of nature	MOLECULAR GENETICS • Reproduction • Inheritance • DNA and mutations • Mendel's work
LIFEPAC 5	ELECTRICITY AND MAGNETISM • Electric current • Electric circuits • Magnetic materials • Electricity and magnets	TRANSFORMATION OF ENERGY • Work and energy • Heat energy • Chemical energy • Energy sources	CHEMICAL STRUCTURE • Nature of matter • Periodic Table • Diagrams of atoms • Acids and bases
LIFEPAC 6	CHANGES IN MATTER • Properties of water • Properties of matter • Molecules and atoms • Elements	RECORDS IN ROCK: THE FLOOD • The Biblical account • Before the flood • The flood • After the flood	LIGHT AND SOUND • Sound waves • Light waves • The visible spectrum • Colors
LIFEPAC 7	WEATHER • Causes of weather • Forces of weather • Observing weather • Weather instruments	RECORDS IN ROCK: FOSSILS • Fossil types • Fossil location • Identifying fossils • Reading fossils	MOTION AND ITS MEASUREMENT • Definition of force • Rate of doing work • Laws of motion • Change in motion
LIFEPAC 8	THE SOLAR SYSTEM • Our solar system • The big universe • Sun and planets • Stars and space	RECORDS IN ROCK: GEOLOGY • Features of the earth • Rock of the earth • Forces of the earth • Changes in the earth	SPACESHIP EARTH • Shape of the earth • Rotation and revolution • Eclipses • The solar system
LIFEPAC 9	THE PLANET EARTH • The atmosphere • The hydrosphere • The lithosphere • Rotation and revolution	CYCLES IN NATURE • Properties of matter • Changes in matter • Natural cycles • God's order	SUN AND OTHER STARS • The sun • Investigating stars • Common stars • Constellations
LIFEPAC 10	GOD'S CREATION • Earth and solar system • Matter and weather • Using nature • Conservation	LOOK AHEAD • Plant and animal life • Balance in nature • Biblical records • Records of rock	THE EARTH AND THE UNIVERSE • Plant systems • Animal systems • Physics and chemistry • The earth and stars

Science LIFEPAC Overview

	Grade 7	Grade 8	Grade 9
LIFEPAC 1	WHAT IS SCIENCE • Tools of a scientist • Methods of a scientist • Work of a scientist • Careers in science	SCIENCE AND SOCIETY • Definition of science • History of science • Science today • Science tomorrow	OUR ATOMIC WORLD • Structure of matter • Radioactivity • Atomic nuclei • Nuclear energy
LIFEPAC 2	PERCEIVING THINGS • History of the metric system • Metric units • Advantages of the metric system • Graphing data	STRUCTURE OF MATTER I • Properties of matter • Chemical properties of matter • Atoms and molecules • Elements, compounds, & mixtures	VOLUME, MASS and DENSITY • Measure of matter • Volume • Mass • Density
LIFEPAC 3	EARTH IN SPACE I • Ancient stargazing • Geocentric Theory • Copernicus • Tools of astronomy	STRUCTURE OF MATTER II • Changes in matter • Acids • Bases • Salts	PHYSICAL GEOLOGY • Earth structures • Weathering and erosion • Sedimentation • Earth movements
LIFEPAC 4	EARTH IN SPACE II • Solar energy • Planets of the sun • The moon • Eclipses	HEALTH AND NUTRITION • Foods and digestion • Diet • Nutritional diseases • Hygiene	HISTORICAL GEOLOGY • Sedimentary rock • Fossils • Crustal changes • Measuring time
LIFEPAC 5	THE ATMOSPHERE • Layers of the atmosphere • Solar effects • Natural cycles • Protecting the atmosphere	ENERGY I • Kinetic and potential energy • Other forms of energy • Energy conversions • Entropy	BODY HEALTH I • Microorganisms • Bacterial infections • Viral infections • Other infections
LIFEPAC 6	WEATHER • Elements of weather • Air masses and clouds • Fronts and storms • Weather forecasting	ENERGY II • Magnetism • Current and static electricity • Using electricity • Energy sources	BODY HEALTH II • Body defense mechanisms • Treating disease • Preventing disease • Community health
LIFEPAC 7	CLIMATE • Climate and weather • Worldwide climate • Regional climate • Local climate	MACHINES I • Measuring distance • Force • Laws of Newton • Work	ASTRONOMY • Extent of the universe • Constellations • Telescopes • Space explorations
LIFEPAC 8	HUMAN ANATOMY I • Cell structure and function • Skeletal and muscle systems • Skin • Nervous system	MACHINES II • Friction • Levers • Wheels and axles • Inclined planes	OCEANOGRAPHY • History of oceanography • Research techniques • Geology of the ocean • Properties of the ocean
LIFEPAC 9	HUMAN ANATOMY II • Respiratory system • Circulatory system • Digestive system • Endocrine system	BALANCE IN NATURE • Photosynthesis • Food • Natural cycles • Balance in nature	SCIENCE AND TOMORROW • The land • Waste and ecology • Industry and energy • New frontiers
LIFEPAC 10	CAREERS IN SCIENCE • Scientists at work • Astronomy • Meteorology • Medicine	SCIENCE AND TECHNOLOGY • Basic science • Physical science • Life science • Vocations in science	SCIENTIFIC APPLICATIONS • Measurement • Practical health • Geology and astronomy • Solving problems

Grade 10	Grade 11	Grade 12	
TAXONOMY • History of taxonomy • Binomial nomenclature • Classification • Taxonomy	INTRODUCTION TO CHEMISTRY • Metric units and instrumentation • Observation and hypothesizing • Scientific notation • Careers in chemistry	KINEMATICS • Scalars and vectors • Length measurement • Acceleration • Fields and models	LIFEPAC 1
BASIS OF LIFE • Elements and molecules • Properties of compounds • Chemical reactions • Organic compounds	BASIC CHEMICAL UNITS • Alchemy • Elements • Compounds • Mixtures	DYNAMICS • Newton's Laws of Motion • Gravity • Circular motion • Kepler's Laws of Motion	LIFEPAC 2
MICROBIOLOGY • The microscope • Protozoan • Algae • Microorganisms	GASES AND MOLES • Kinetic theory • Gas laws • Combined gas law • Moles	WORK AND ENERGY • Mechanical energy • Conservation of energy • Power and efficiency • Heat energy	LIFEPAC 3
CELLS • Cell theories • Examination of the cell • Cell design • Cells in organisms	ATOMIC MODELS • Historical models • Modern atomic structure • Periodic Law • Nuclear reactions	WAVES • Energy transfers • Reflection and refraction of waves • Diffraction and interference • Sound waves	LIFEPAC 4
PLANTS: GREEN FACTORIES • The plant cell • Anatomy of the plant • Growth and function of plants • Plants and people	CHEMICAL FORMULAS • Ionic charges • Electronegativity • Chemical bonds • Molecular shape	LIGHT • Speed of light • Mirrors • Lenses • Models of light	LIFEPAC 5
HUMAN ANATOMY AND PHYSIOLOGY • Digestive and excretory system • Respiratory and circulatory system • Skeletal and muscular system • Body control systems	CHEMICAL REACTIONS • Detecting reactions • Energy changes • Reaction rates • Equilibriums	STATIC ELECTRICITY • Nature of charges • Transfer of charges • Electric fields • Electric potential	LIFEPAC 6
INHERITANCE • Gregor Mendel's experiments • Chromosomes and heredity • Molecular genetics • Human genetics	EQUILIBRIUM SYSTEMS • Solutions • Solubility equilibriums • Acid-base equilibriums • Redox equilibriums	CURRENT ELECTRICITY • Electromotive force • Electron flow • Resistance • Circuits	LIFEPAC 7
CELL DIVISION & REPRODUCTION • Mitosis and meiosis • Asexual reproduction • Sexual reproduction • Plant reproduction	HYDROCARBONS • Organic compounds • Carbon atoms • Carbon bonds • Saturated and unsaturated	MAGNETISM • Fields • Forces • Electromagnetism • Electron beams	LIFEPAC 8
ECOLOGY & ENERGY • Ecosystems • Communities and habitats • Pollution • Energy	CARBON CHEMISTRY • Saturated and unsaturated • Reaction types • Oxygen groups • Nitrogen groups	ATOMIC AND NUCLEAR PHYSICS • Electromagnetic radiation • Quantum theory • Nuclear theory • Nuclear reaction	LIFEPAC 9
APPLICATIONS OF BIOLOGY • Principles of experimentation • Principles of reproduction • Principles of life • Principles of ecology	ATOMS TO HYDROCARBONS • Atoms and molecules • Chemical bonding • Chemical systems • Organic chemistry	KINEMATICS TO NUCLEAR PHYSICS • Mechanics • Wave motion • Electricity • Modern physics	LIFEPAC 10

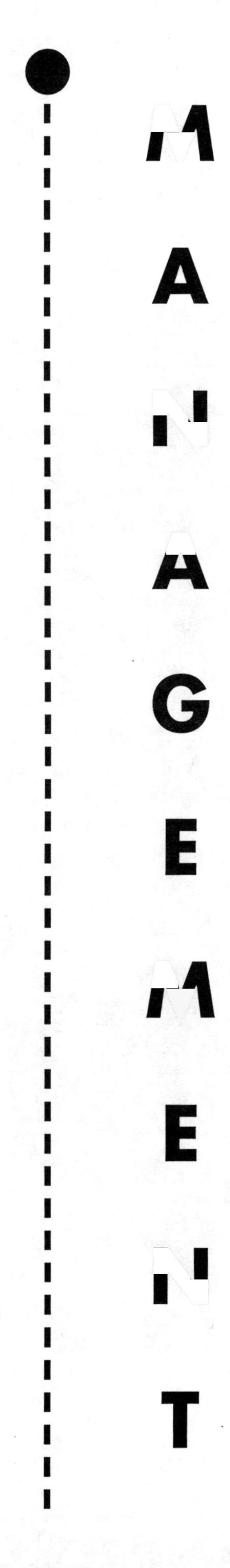
LIFEPAC
MANAGEMENT

STRUCTURE OF THE LIFEPAC CURRICULUM

The LIFEPAC curriculum is conveniently structured to provide one teacher's guide containing teacher support material with answer keys and ten student worktexts for each subject at grade levels two through twelve. The worktext format of the LIFEPACs allows the student to read the textual information and complete workbook activities all in the same booklet. The easy to follow LIFEPAC numbering system lists the grade as the first number(s) and the last two digits as the number of the series. For example, the Language Arts LIFEPAC at the 6th grade level, 5th book in the series would be LA 605.

Each LIFEPAC is divided into 2 to 5 sections and begins with an introduction or overview of the booklet as well as a series of specific learning objectives to give a purpose to the study of the LIFEPAC. The introduction and objectives are followed by a vocabulary section which may be found at the beginning of each section at the lower levels, at the beginning of the LIFEPAC in the middle grades, or in the glossary at the high school level. Vocabulary words are used to develop word recognition and should not be confused with the spelling words introduced later in the LIFEPAC. The student should learn all vocabulary words before working the LIFEPAC sections to improve comprehension, retention and reading skills.

Each activity or written assignment has a number for easy identification. The first digit is the section number and the digit(s) to the right of the decimal is the number of the activity.

Teacher checkpoints which are essential to maintain quality learning are found at various locations throughout the LIFEPAC. The teacher should check for 1) neatness of work and penmanship, 2) quality of understanding (checked with a short oral quiz), 3) thoroughness of preceding answers (complete sentences and paragraphs, correct spelling, etc.), 4) all activities being attempted (no blank spaces), 5) accuracy of answers as compared to the answer key (all answers correct).

The self test questions are also number coded for easy reference. The first digit is the section number, the zero indicates that it is a self test question, and the digits to the right of the zero are the question number. For example, 2.015 means that this is the 15th question in the self test of Section II.

The LIFEPAC test is packaged at the centerfold of each LIFEPAC. It should be removed and put aside before giving the booklet to the student for study.

Answer and test keys have the same numbering system as the LIFEPACs and appear at the back of this guide. The student may have access to the answer keys (not the test key) under teacher supervision so he can score his own work.

A thorough study of the Curriculum Overview by the teacher before instruction begins is essential to the success of the student. The teacher should become familiar with expected skill mastery and understand how these grade level skills fit into the overall skill development of the curriculum. The teacher should also preview the objectives that appear at the beginning of each LIFEPAC for additional preparation and planning.

TEST SCORING and GRADING

Answer keys and test keys give examples of correct answers. They convey the idea but the student may use many ways to express a correct answer. The teacher should check for the essence of the answer, not for the exact wording. Many questions are high level and require thinking and creativity on the part of the student. Each answer should be scored based on whether or not the main idea written by the student matches the model example. "Any Order" or "Either Order" in a key indicates that no particular order is necessary to be correct.

Most self tests and LIFEPAC tests at the lower elementary levels are scored at 1 point per answer; however, the upper levels may have a point system awarding 2 to 5 points for various answers or questions. At any level the total test points will vary. They may be 78, 85, 100, 105, etc., that is, they may not always be 100.

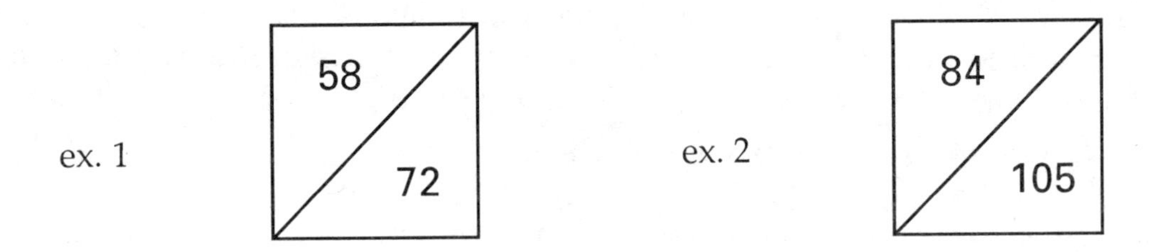

A score box similar to ex.1 above should appear at the end of each self test and LIFEPAC test. The bottom score, 72, represents the total number of points possible on this test. The upper score, 58, represents the number of points your student will need to receive an 80% or passing grade. If you wish to establish the exact percentage that your student has achieved, find the total points of his correct answers and divide it by the bottom number (in this case 72.) For example, if your student has a point total of 65, divide 65 by 72 for a grade of 90%. Referring to ex. 2, on a test with a total of 105 possible points, the student would have to receive a minimum of 84 correct points for an 80% or passing grade. If your student has received 93 points, simply divide the 93 by 105 for a percentage grade of 89%. Students who receive a score below 80% should review the LIFEPAC and retest using the appropriate Alternate Test found in the teacher's guide.

The following is a guideline to assign letter grades for completed LIFEPACs based on a maximum total score of 100 points.

LIFEPAC Test	=	60% of the Total Score (or percent grade)
Self Test	=	25% of the Total Score (average percent of self tests)
Reports	=	10% or 10* points per LIFEPAC
Oral Work	=	5% or 5* points per LIFEPAC

*Determined by the teacher's subjective evaluation of the student's daily work.

Example:

LIFEPAC Test Score	=	92%	92 x .60	=	55 points
Self Test Average	=	90%	90 x .25	=	23 points
Reports				=	8 points
Oral Work				=	4 points
TOTAL POINTS				=	90 points

Grade Scale based on point system:

100	–	94	=	A
93	–	86	=	B
85	–	77	=	C
76	–	70	=	D
Below		70	=	F

TEACHER HINTS and STUDYING TECHNIQUES

LIFEPAC Activities are written to check the level of understanding of the preceding text. The student may look back to the text as necessary to complete these activities; however, a student should never attempt to do the activities without reading (studying) the text first. Self tests and LIFEPAC tests are never open book tests.

Language arts activities (skill integration) often appear within other subject curriculum. The purpose is to give the student an opportunity to test his skill mastery outside of the context in which it was presented.

Writing complete answers (paragraphs) to some questions is an integral part of the LIFEPAC Curriculum in all subjects. This builds communication and organization skills, increases understanding and retention of ideas, and helps enforce good penmanship. Complete sentences should be encouraged for this type of activity. Obviously, single words or phrases do not meet the intent of the activity, since multiple lines are given for the response.

Review is essential to success. Time invested in review where review is suggested will be time saved in correcting errors later. Self tests, unlike the section activities, are closed book. This procedure helps to identify weaknesses before they become too great to overcome. Certain objectives from self tests are cumulative and test previous sections; therefore, good preparation for a self test must include all material studied up to that testing point.

The following procedure checklist has been found to be successful in developing good study habits in the LIFEPAC curriculum.

1. Read the introduction and Table of Contents.
2. Read the objectives.
3. Recite and study the entire vocabulary (glossary) list.
4. Study each section as follows:
    a. Read the introduction and study the section objectives.
    b. Read all the text for the entire section, but answer none of the activities.
    c. Return to the beginning of the section and memorize each vocabulary word and definition.
    d. Read the section, answer activities, check the answers with the answer key, correct all errors and do the adult checks.
    e. Read the self test but do not answer the questions.
    f. Go to the beginning of the first section and reread the text and answers to the activities up to the self test you have not yet done.
    g. Answer the questions to the self test without looking back.
    h. Have the self test checked by the teacher.
    i. Correct the self test along with a adult check.
    j. Repeat steps a–i for each section.

5. Use the SQ3R* method to prepare for the LIFEPAC test.
6. Take the LIFEPAC test as a closed book test.
7. LIFEPAC tests are administered and scored under direct teacher supervision. Students who receive scores below 80% should review the LIFEPAC using the SQ3R* study method and take the Alternate Test located in the Teacher's Guide. The final test grade may be the grade on the Alternate Test or an average of the grades from the original LIFEPAC test and the Alternate Test.

*SQ3R: Scan the whole LIFEPAC.
Question yourself on the objectives.
Read the whole LIFEPAC again.
Recite through an oral examination.
Review weak areas.

GOAL SETTING and SCHEDULES

Each school must develop its own schedule, because no one set of procedures will fit every situation. The following is an example of a daily schedule that includes the five subjects as well as time slotted for special activities.

Possible Daily Schedule

8:15	–	8:25	Pledges, prayer, songs, devotions, etc.
8:25	–	9:10	Bible
9:10	–	9:55	Language Arts
9:55	–	10:15	Recess (juice break)
10:15	–	11:00	Mathematics
11:00	–	11:45	Social Studies
11:45	–	12:30	Lunch, recess, quiet time
12:30	–	1:15	Science
1:15	–		Drill, remedial work, enrichment*

*Enrichment: Computer time, physical education, field trips, fun reading, games and puzzles, family business, hobbies, resource persons, guests, crafts, creative work, electives, music appreciation, projects.

Basically, two factors need to be considered when assigning work to a student in the LIFEPAC curriculum.

The first is time. An average of 45 minutes should be devoted to each subject, each day. Remember, this is only an average. Because of extenuating circumstances a student may spend only 15 minutes on a subject one day and the next day spend 90 minutes on the same subject.

The second factor is the number of pages to be worked in each subject. A single LIFEPAC is designed to take 3 to 4 weeks to complete. Allowing about 3-4 days for LIFEPAC introduction, review, and tests, the student has approximately 15 days to complete the LIFEPAC pages. Simply take the number of pages in the LIFEPAC, divide it by 15 and you will have the number of pages that must be completed on a daily basis to keep the student on schedule. For example, a LIFEPAC containing 45 pages will require 3 completed pages per day. Again, this is only an average. While working a 45 page LIFEPAC, the student may complete only 1 page the first day if the text has a lot of activities or reports but go on to complete 5 pages the next day.

Long range planning requires some organization. Because the traditional school year originates in the early fall of one year and continues to late spring of the following year, a calendar should be devised that covers this period of time. Approximate beginning and completion dates can be noted on the calendar as well as special occasions such as holidays, vacations and birthdays. Since each LIFEPAC takes 3-4 weeks or eighteen days to complete, it should take about 180 school days to finish a set of ten LIFEPACs. Starting at the beginning school date, mark off eighteen school days on the calendar and that will become the targeted completion date for the first LIFEPAC. Continue marking the calendar until you have established dates for the remaining nine LIFEPACs making adjustments for previously noted holidays and vacations. If all five subjects are being used, the ten established target dates should be the same for the LIFEPACs in each subject.

FORMS

The sample weekly lesson plan, and student grading sheet, are included in this section as teacher support materials and may be duplicated at the convenience of the teacher.

The student grading sheet is provided for those who desire to follow the suggested guidelines for assignment of letter grades found on pages 14-15 of this section. The student's self test scores should be posted as percentage grades. When the LIFEPAC is completed the teacher should average the self test grades, multiply the average by .25 and post the points in the box marked self test points. The LIFEPAC percentage grade should be multiplied by .60 and posted. Next, the teacher should award and post points for written reports and oral work. A report may be any type of written work assigned to the student whether it is a LIFEPAC or additional learning activity. Oral work includes the student's ability to respond orally to questions which may or may not be related to LIFEPAC activities or any type of oral report assigned by the teacher. The points may then be totaled and a final grade entered along with the date that the LIFEPAC was completed.

The Student Record Book which was specifically designed for use with the Alpha Omega curriculum provides space to record weekly progress for one student over a nine week period as well as a place to post self test and LIFEPAC scores. The Student Record Books are available through the current Alpha Omega catalog; however, unlike the enclosed forms these books are not for duplication and should be purchased in sets of four to cover a full academic year.

WEEKLY LESSON PLANNER				
			Week of:	
	Subject	Subject	Subject	Subject
Monday				
	Subject	Subject	Subject	Subject
Tuesday				
	Subject	Subject	Subject	Subject
Wednesday				
	Subject	Subject	Subject	Subject
Thursday				
	Subject	Subject	Subject	Subject
Friday				

WEEKLY LESSON PLANNER

			Week of:
Subject	Subject	Subject	Subject
Monday			
Subject	Subject	Subject	Subject
Tuesday			
Subject	Subject	Subject	Subject
Wednesday			
Subject	Subject	Subject	Subject
Thursday			
Subject	Subject	Subject	Subject
Friday			

Student Name ______________________ Year ____________

Bible

LP #	Self Test Scores by Sections 1	2	3	4	5	Self Test Points	LIFEPAC Test	Oral Points	Report Points	Final Grade	Date
01											
02											
03											
04											
05											
06											
07											
08											
09											
10											

History & Geography

LP #	Self Test Scores by Sections 1	2	3	4	5	Self Test Points	LIFEPAC Test	Oral Points	Report Points	Final Grade	Date
01											
02											
03											
04											
05											
06											
07											
08											
09											
10											

Language Arts

LP #	Self Test Scores by Sections 1	2	3	4	5	Self Test Points	LIFEPAC Test	Oral Points	Report Points	Final Grade	Date
01											
02											
03											
04											
05											
06											
07											
08											
09											
10											

Student Name ____________________ Year __________

Mathematics

LP #	Self Test Scores by Sections 1	2	3	4	5	Self Test Points	LIFEPAC Test	Oral Points	Report Points	Final Grade	Date
01											
02											
03											
04											
05											
06											
07											
08											
09											
10											

Science

LP #	Self Test Scores by Sections 1	2	3	4	5	Self Test Points	LIFEPAC Test	Oral Points	Report Points	Final Grade	Date
01											
02											
03											
04											
05											
06											
07											
08											
09											
10											

Spelling/Electives

LP #	Self Test Scores by Sections 1	2	3	4	5	Self Test Points	LIFEPAC Test	Oral Points	Report Points	Final Grade	Date
01											
02											
03											
04											
05											
06											
07											
08											
09											
10											

NOTES

INSTRUCTIONS FOR SCIENCE

The LIFEPAC curriculum for grades two through twelve was written with the daily instructional material written directly in the LIFEPACs. The student is encouraged to read and follow his own instructional material, thus developing independent study habits. The teacher should introduce the LIFEPAC to the student, set a required completion schedule, complete adult checks, be available for questions regarding both subject content and procedures, administer and grade tests, and develop additional learning activities as desired. Teachers working with several students may schedule their time so that students are assigned a quiet work activity when it is necessary to spend instructional time with one particular student.

The Teacher Notes section of the Teacher's Guide lists the required or suggested materials for the LIFEPACs and provides additional learning activities for the students. The materials section refers only to LIFEPAC materials and does not include materials which may be needed for the additional activities. Additional learning activities provide a change from the daily school routine, encourage the student's interest in learning, and may be used as a reward for good study habits.

If you have limited facilities and are not able to perform all the experiments contained in the LIFEPAC curriculum, the Science Project List for grades 3-12 may be a useful tool for you. This list prioritizes experiments into three categories: those essential to perform, those which should be performed as time and facilities permit, and those not essential for mastery of LIFEPACs. Of course, for complete understanding of concepts and student participation in the curriculum, all experiments should be performed whenever practical. Materials for the experiments are shown in Teacher Notes - Materials Needed.

Science Projects List

Key

(1) = Those essential to perform for basic understanding of scientific principles.

(2) = Those which should be performed as time permits.

(3) = Those not essential for mastery of LIFEPACs.

S = Equipment needed for home school or Christian school lab.

E = Explanation or demonstration by instructor may replace student or class lab work.

H = Suitable for homework or for home school students. (No lab equipment needed.)

Science 501

501.A	(1)	S
501.B	(1)	S
501.C	(1)	S
501.D	(1)	S
501.E	(1)	S

Science 502

502.A	(1)	S
502.B	(2)	H
Seed Hunt	(1)	H
Cones	(2)	H
502.C	(1)	S

Science 503

503.A	(1)	S
503.B	(1)	S

Science 504

504	(2)	H

Science 505

505.A	(1)	S
505.B	(3)	H
505.C	(1)	S
505.D	(3)	S

Science 506

None

Science 507

507.A	(1)	S
507.B	(1)	S

Science 508

508.A	(2)	H or S
508.B	(1)	S
Rock Hunt	(1)	H
508.C	(2)	H

Science 509

509.A	(2)	H
509.B	(3)	H
509.C	(1)	H
509.D	(1)	H
509.E	(2)	S
509.F	(1)	S
509.G	(3)	E

Science 510

Rocks	(1)	S

Materials Needed for LIFEPAC

Required:

- optical microscope
- slides
- slide covers
- toothpicks
- diluted iodine solution
- 3 small jars half full of water
- onion bulb
- knife or scalpel
- tweezers
- small eyedropper
- sterile needle
- ink stain
- pond water
- cotton ball
- rubbing alcohol

Suggested:

None

Additional Learning Activities

Section I The Basic Unit of Living Things: A Cell

1. Introduce students to the use of a microscope. Show them how to magnify and focus. Have students look at a drop of water, hair, paper, thread, and so forth for practice.
2. Students make drawings of a cell and label the nucleus, membrane, and cytoplasm. Under the drawing list the three parts and write a one-sentence description for each.
3. Groups of students prepare slides of several different fruits and vegetables. Observe and discuss your slides.
4. Look up information on Robert Hooke and write a short report.
5. Be creative! Write a few paragraphs and illustrate a story titled: "A cell Named ____(your name)__________." Include your parts and functions.

Section II The Life and Activity of Cells

1. Have two green stalks of celery, one firm and one wilted. Discuss their color and rigidity. Encourage students to study Section II to find out why stalk #2 is weak and wilted.
2. The class divides into pairs. Each student draws five different kinds of cells that they studied in Section II. Students exchange papers and label their partner's drawing.
3. Make a word puzzle like the one in Section II. Students can exchange puzzles and write the meanings of the words they unscrambled.
4. Take two stalks of celery. Put each stalk in a glass of water. Add red or blue food coloring to one glass of water. Observe the glasses the next day. Write a report on what you observed. See if you can explain the results.
5. Draw pictures of the two kinds of blood cells and explain their functions.

Section III Energy and Growth of Cells

1. Take two plants. Put one in the sunlight and one in the dark. Check the plants after one week. What happened? Why?
2. Cut out felt shapes for the sun, a few animals, and a few plants. Use the felt pieces to explain the carbon cycle in your own words. Do the same for photosynthesis.
3. Make your own drawings to explain mitosis. Show your drawings to a friend or the class and explain the process.
4. Observe or read about a sunflower. Write a few sentences explaining why you think the sunflower follows the sun.

Materials Needed for LIFEPAC

Required:

- two flowers
- microscope
- slides
- slide covers
- magnifying glass or dissection scope
- straight pins
- paper
- dropper
- sharp knife
- slice of bread
- large jar with lid
- tweezers
- lima bean seed
- corn seed
- small jar half full of water
- male and female pine cones

Suggested:

None

Additional Learning Activities

Section I Classifying Living Things and Plants

1. On the chalkboard draw a picture of the life cycle of a corn plant. Introduce the four life stages.
2. Search the library, magazines, or Internet for pictures of living things from each of the 5 kingdoms of living things.

Section II Seed-Bearing Plants

1. If possible, show students a flower. Demonstrate the activity in Experiment 502.A while you explain the reproduction of flowering plants. Show example pictures to students in class.
2. Show a male and female cone. Explain the way cones reproduce.
3. Tell students that you have shown two ways that plants reproduce. They should read Section I to learn more about flowering plants, cones, and another type of plant reproduction called vegetative reproduction.
4. Place some radish seeds between layers of moist paper towels and set them in a warm place. After several days, examine the root hairs with a hand lens. You can show that roots "seek" water. Put some radish seeds or grass seeds on blotting paper or folded paper towels. Place the seeds and paper between two glass plates. Bind the setup with rubber bands and stand it on edge in a shallow pan of water. After the roots have grown about a half inch, notice the direction in which the roots are growing. Give the glass plates a quarter turn. After a few days notice the new direction in which the roots are growing. Continue the procedure until a full turn has been made.

Section III Spore-Bearing Plants and Fungi

1. Introduce the vocabulary word *spore*. Explain that spore-bearing plants do not have flowers or cones but reproduce by making spores. Show pictures or draw examples of ferns, mushrooms, and mold.
2. Students fold a paper in half. On one half illustrate the life cycle of one flowering plant. Illustrate the life cycle of one spore-bearing plants on the other side. Show the drawings to a friend and explain the life cycles.
3. Look up information on mushrooms. Find out why people should be very careful when picking mushrooms to eat. Report your findings to the class.

Section IV One-Celled Living Things

1. Show students a picture of algae floating on top of a pond. Explain that the green film is composed of many, many one-celled organisms.
2. Hold up a loaf of bread. Ask students if they know what makes the bread rise. Explain that yeast like algae is also a one-celled organism that reproduces very rapidly.
3. Dissolve a spoonful of sugar in a glass of water. Drop a small piece from a cake of yeast into the glass. Place the glass in the sunlight. Do you see bubbles rising from the piece of yeast? What process is going on? Of what are the bubbles made? Write a report of your observations, including your answers to these questions. Share your report with the class.

Materials Needed for LIFEPAC

Required:

- baby food jar or Mason jar
- bran flakes or oatmeal
- potato
- mealworm
- knife
- chicken egg
- dropper
- small dish
- tissue paper
- magnifying glass
- food coloring

Suggested:

- None

Additional Learning Activities

Section I Invertebrates

1. Groups of four or five students work together to make a chart classifying invertebrates. Headings should be One-Celled Animal-like Protists, Insects, Worms, Mollusks. Include all invertebrates mentioned in Section I plus any more you know or wish to look up in the library.
2. Look up more information about butterflies. Draw pictures and explain the life cycle of one kind of butterfly.

Section II Vertebrates

1. Students may cut out animal pictures from magazines. Each student may show one picture, name the animal, and tell whether it is a vertebrate or an invertebrate.
2. Make a large chart to classify vertebrates. One student will show a picture of a vertebrate and tell whether it is a fish, amphibian, reptile, bird, or mammal. The student will write the animal name in the appropriate section of the chart. Continue with other students.
3. Ask your parents to take you to the zoo. List all the animals you see. Next to each animal name write *vertebrate* or *invertebrate*. Tell whether it is a worm, mollusk, reptile, and so forth. Example: alligator-vertebrate-reptile.
4. If you live near a pond, try to catch some tadpoles. Watch their growth. Draw pictures of what you observe.

Materials Needed for LIFEPAC

Required:
None

Suggested:
large container (glass or clear plastic) with cover or lid
gravel
several cans or small jars
2 liters potting soil
small shovel
sand
bottle cap
a few small plants
a few small animals (insects, frogs, worms, etc.)
small sprinkling can

Additional Learning Activities

Section I The Balance of Nature

1. Use food chain drawings in Section I for this activity. Have one student explain his drawing of a food chain. Have a second student explain what would happen if the producer were removed. A third student could explain why the primary consumer is necessary and what would happen if it were removed. Continue until each part of the food chain has been explained.
2. Demonstrate a water cycle. Students can bring margarine tubs, add soil, add water and cover. Make observations the next day. Students explain what they saw.

Section II The Prairie Web of Life

1. Draw a "Web of Life" for the prairie. Include as many plants and animals as possible. Show it to the class. Explain what happens when a drought occurs or when there are heavy rains.
2. Write a story about the prairie. Pretend you were with the first group of pioneers to settle in the prairie. Tell what you think life was like for the first settlers. (Laura Ingalls Wilder books or the computer game, The Oregon Trail, would be good reference material).

Section III Humans and the Web of Life

1. One student can lead the group in this activity. Have students fold a paper in half. Have them write the heading "Pollution" on one half, and "Stewardship" on the other half. They should list five ways man has polluted the environment on one side and five ways they can be good stewards on the other side.
2. Draw three food chains using man as:
    1. producer
    2. primary consumer
    3. secondary consumer
3. Write a report explaining how people can change the balance of nature.
4. Find out what the hunting laws and seasons are in your area. Write a few paragraphs telling whether you think these laws are good ones or not. Explain the reasons for your opinion.

Materials Needed for LIFEPAC

Required:

- yellow construction paper
- thermometer
- 3 liquid ounces of vinegar
- one tablespoon of baking soda
- one glass jar
- two paper clips
- water
- small jar or baby-food jar

Suggested:

None

ADDITIONAL LEARNING ACTIVITIES

Section I Energy and Work

1. Students make a list of all types of energy discussed in Section I (chemical, heat, mechanical, light, electrical, and sound). Have them write an example of each.
2. Students find magazine pictures of work being done. Students could label pictures according to the type of energy being used. Example: a picture of an explosion would be labeled *sound energy.*
3. Prepare a demonstration of one type of energy that you studied.
4. Demonstrate potential, kinetic, and mechanical energy to a friend.

Section II Work from Energy

1. Five or six students bring in flashlights. Divide the class into groups. Take out the batteries, examine them, and demonstrate how the chemical energy from the batteries can do work (make the light shine). Decide whether any heat energy is produced.
2. Find a good book of science experiments. Have your parents help you choose one experiment using a battery to do work.
3. Make a diagram explaining how gasoline makes a car engine work.

Section III Energy in the Future

1. Students may read and discuss their surveys from Section III on how to help solve energy problems.
2. Invite the owner of a local gas station to visit the class and talk about why there is an energy crisis and what can be done about it.
3. Look up information on solar heating. Draw a diagram of a simple solar heater and explain it to the class.
4. Look up information on nuclear power. Write a report telling about the benefits and dangers of nuclear energy.

Materials Needed for LIFEPAC

Required:
None

Suggested:
None

ADDITIONAL LEARNING ACTIVITIES

Section I Before The Flood

1. Read selected parts of Genesis, Chapters 1 through 3 to the class.
2. Involve the whole class in making a chart to compare facts given in the Bible to scientific data about the earth before the Flood.
3. Half of the class may make a mural of life before the Flood as described in the Bible. The other half of the class makes a mural of life before the Flood as recorded by scientists.
4. Do research on dinosaurs. Name two or more kinds of dinosaurs. Tell where they lived, what they ate, and how big they were. Illustrate.
5. Look up information on fossils. Find out how fossils can tell a story.

Section II The Flood

1. Students may add a section to the murals started after Section I. Name it "The Flood."
2. Make a fossil! You can use modeling clay or play dough. Bring a bone, rock, leaf, or shell to school. Smooth out some clay. Press an object into it. Carefully lift out the object. Let the clay harden by being exposed to the air. Examine your fossil with a magnifying glass as scientists do.
3. Look up information about coal. Write a report telling how coal was formed, where to find it, and how it is used. Use an empty aquarium or large glass bowl to show the formation to the class. (You will need layers of mud, peat, plants, and water.)
4. Make a model of the Ark. Choose materials to resemble the actual ones used. Show your model and explain how Noah made the real Ark.

Section III After The Flood

1. Add a third section to the two murals started after Section I. Have students review Section III and include many details. Name it "After the Flood."
2. Make a relief map using papier-mâché. Plan out areas to show volcanoes, earthquakes, glaciers, oceans, and so forth. Paint the finished product and label.
3. Find out more about volcanoes. Where are volcanoes found today? Why do they erupt? What is lava? What kinds of rocks are formed from lava? Do the activity on glaciers or earthquakes.
4. Find a good book about rocks. Make a rock collection. Identify and label the rocks.

Materials Needed for LIFEPAC

Required:
- small plastic containers (10 cm across)
- modeling clay
- seashells or bones
- molding plaster
- jars for water
- tin cans
- stick

Suggested:
- None

ADDITIONAL LEARNING ACTIVITIES

Section I Fossil Formation

1. Go on a fossil hunt! The class may plan where to look. Take magnifying glasses and see what you can find.
2. Find several different kinds of fossils. Follow the directions given in Section I. Label your fossils and make a display.

Section II Reading Fossils

1. Show a picture of a very complete fossil or reconstructed skeleton. Ask students to look for clues, make inferences, and come to a conclusion about the identity of the picture.
2. If possible, have several fossils available for students to examine. If not, use pictures. Students identify fossils as plant or animal. If a plant, tell whether they are:
   a. seed-bearing,
   b. spore-bearing, or
   c. single-celled

   If animal, tell whether they are:
   a. Vertebrate:
      1. fish,
      2. amphibian,
      3. reptile,
      4. bird, or
      5. mammal

   b. Invertebrate:
      1. insect,
      2. worm, or
      3. mollusk

   Example: Shell fossil-animal-invertebrate-mollusk
3. Use your imagination to draw a reconstructed skeleton, or a model of a prehistoric animal. On the other side of the paper, draw what you think the animal should look like. Color your drawing.
4. Imagine you lived in prehistoric times. Write a short story telling about your home, pets, plants, foods, and favorite things to do.

Materials Needed for LIFEPAC

Required:

- table salt
- colored paper
- magnifying glass
- mixture of sand, soil, large and small pebbles
- glass of water
- orange
- plastic tub or metal pan
- short board (about 50 cm long)

Suggested: None

ADDITIONAL LEARNING ACTIVITIES

Section I Structure of the Earth

1. Show the class a world globe. Explain that the surface of the earth includes all the areas we can see.
2. Show the poles and explain the spherical shape of the earth.
3. Ask the students to name and point to different landforms they recognize (ocean, mountain, etc.).
4. Rock identification tests: Divide the class into groups of four. Give each group five different kinds of rocks. Have students test for:
   a. color – by observation
   b. luster – dull, medium, or bright by observation
   c. streak test – rub rocks against a cement walk
   d. hardness – use a penny to scratch rocks
   e. cleavage – by observation

   Have students record their observations on a chart.

Section II Changes in the Earth

1. Pantomime the ways that people change the earth. Have other students guess the way.
2. Make a volcano. Use colored clay for the layers of rock. Put a very small container of vinegar at the bottom. To cause eruption, drop some baking soda down the passage way.
3. Look up information on the San Andreas Fault in California. Write a report explaining why scientists think a major earthquake will soon occur there.

Materials Needed for LIFEPAC

Required:

- four differently shaped glass containers (goblets, tall jars, and so forth)
- baby-food jar
- candle in holder
- matches
- large glass plate or pan
- globe with tilted axis
- tall, thin jar or a graduated cylinder
- small marble
- large marble
- masking tape
- toothpick
- 2 glass containers of water
- ruler with centimeter markings
- spoonful of sugar
- equal arm balance
- two baby-food jars and lids
- ice cubes
- droppers
- liquid measure
- pair of identical pans
- spoon or stirrer
- funnel

Suggested:

None

ADDITIONAL LEARNING ACTIVITIES

Section I Matter

1. Use your Science LIFEPAC 509 to introduce and demonstrate properties of matter. Measure its volume. Weigh its density. Talk about its special properties. Discuss how it could be physically or chemically changed.
2. Show pictures or a model of a molecule of water. Discuss atoms and how their combinations determine what kind of matter is formed.
3. Each student brings four objects to school. Students make a chart of the physical properties of this matter. Headings should be:
   - Matter (Write name or draw picture of the object.)
   - Density (Use a scale to weigh objects or use feel. Number 1, 2, 3, 4; 1 = most dense to 4 = least dense.)
   - Volume (Use displacement, or number objects 1, 2, 3, 4. Use 1 for most volume to 4 for least. You also could measure height, width, and length to figure volume.)
   - Special properties (stretch, brittleness, and so forth.)
4. Make some popcorn. If possible, a glass, or clear plastic popper would be perfect. First, weigh the oil and popcorn kernels before placing them in the popper. Second, have students observe the popcorn popping. (Ask how they could compare what they see with the action of molecules.) Finally, weigh the popcorn when it is finished popping. Have students draw conclusions about density (before and after), volume (before and after), gain or loss of matter, and if the change was physical or chemical.

5. Think of something you and your mom or dad can make in the kitchen that involves a chemical change. Bring it to school and explain what you did to make a chemical change.
6. Use tinker toys to make models of molecules. Look up information on molecules. Find out the shape of a water molecule, sugar molecule, salt molecule, and so forth. Make models of three or four types of molecules, label them, and show them to the class.

Section II Other Natural Cycles

1. Use a tilted axis globe and a flashlight to demonstrate the change of seasons. Explain that seasons are caused by the movement of the earth. Darken the room and use the flashlight for the sun. Rotate the globe to show each season.
2. Ask students to explain how the seasons could be called a cycle.
3. Review cycles of living things, water cycle, carbon cycle, and mineral cycle with students. Use chalkboard drawings and arrows to show each cycle.
4. Introduce students to Section II to find out about cycles of comets.
5. Students may draw the cycle of a comet. They can label the parts and write two or three sentences to explain the cycle.
6. Students may read their reports on scientists who discovered comets. Make a large chart with the headings: Person, Comet, Year Discovered, Special Properties. As each student gives his report, have other students take turns filling in the chart.
7. Draw the cycle for a *day*.
8. Make a sundial and explain how you can tell time by the sun.

Section III God's Order

1. This section of the LIFEPAC is only a few paragraphs long. Read and discuss it with the class.
2. Read some of the Bible passages suggested in this section. Discuss their meaning with the class.

Materials Needed for LIFEPAC

Required:
- teacher supply of different types of rocks
- newspapers
- magazines
- poster paper
- glue

Suggested:
- None

ADDITIONAL LEARNING ACTIVITIES

Section I Living Things

1. Choose an animal to study. Draw the life cycle of the animal. Tell where it lives, what it eats, and why you chose it.
2. Make a list of three things you can do to be a good steward. Write the days of the week across the top of your list. Check each day to see that you have been a good steward.

Section II The Earth

Make up a puzzle like the one in Section II. Use rock names for the words.

Section III Order in Creation

1. Each student may bring one object to school. Students can show their objects and tell how they can do work.
2. Divide the class into small groups. Assign each group a cycle (water, carbon, seasons, day, etc.). Have the groups plan to draw or dramatize their cycles and explain them.
3. Draw an example of energy being transformed.

ALTERNATE

TESTS

Reproducible Tests
for use with the Science 500
Teacher's Guide

Name ____________________

Write *true* or *false* (each answer, 2 points).

1. _____ Cells are the basic unit of life.
2. _____ All cells have a cell wall.
3. _____ Prokaryote cells have only two parts — an outer membrane and inner protoplasm.
4. _____ Almost all cells are microscopic.
5. _____ Scientists are still making new discoveries today about cells.
6. _____ The cell membrane consists of a phospholipid double layer and proteins.
7. _____ The cell wall is usually made up mostly of water.
8. _____ Photosynthesis in plants produces carbon dioxide and water.
9. _____ Protozoa are an example of a multicellular organism.
10. _____ Budding is a process of cell reproduction.

Complete these statements (each answer, 3 points).

11. Photosynthesis works when a. ____________________, b. ____________________ and c. ____________________ are present in the plant.

12. The carbon cycle needs a. ____________________ and b. ____________________ from plants and c. ____________________ from animals.

13. Red blood cells take a. ____________________ to the cells and b. ____________________ away from them.

14. Psalms 139:14 says that we are a. ____________________ and b. ____________________ made.

Draw and label an animal cell (this answer, 5 points).

15.

Write the correct answer on each line (each answer, 2 points).

16. A ______________________ is needed to see most cells.

    a. flashlight   b. telescope   c. microscope

17. Most multicelled plants and animals reproduce through ____________________________.

    a. budding   b. male-female reproduction c. fusion

18. Chloroplasts are a category of __________________ found in the cells of green plants.

    a. organelles   b. bacteria   c. tissues

19. The splitting apart of cells is know as ________________________.

    a. cell division   b. kenosis   c. oxidation

20. The _______________ in your body determines your hair and eye color.

    a. red blood cell   b. DNA   c. nitrogen

Answer these questions (each answer, 3 points).

21. What are three types of animal tissues?

    a. ___________________________

    b. ___________________________

    c. ___________________________

22. What are the two substances produced by photosynthesis in plants?

a. ______________________________

b. ______________________________

Match these items (each answer, 2 points).

23. _____ cells that fight disease
24. _____ nucleolus
25. _____ energy
26. _____ mitosis
27. _____ xylem
28. _____ respiration
29. _____ cytoplasm
30. _____ God
31. _____ Robert Hooke
32. _____ cell wall

a. the capacity to do work
b. connective tissue in plants
c. fluid material within cell membrane and outside the nucleus
d. discovered cells while looking at cork
e. white blood cells
f. contained within the nucleus of a cell
g. a single cell splits to form two new cells
h. carbon dioxide and energy given off
i. created all living things
j. invented the telescope
k. made of chloroplasts
l. usually made of cellulose

Date ______________________

Score ______________________

Possible Score ______100______

Science 502 Alternate Test

Name ______________________

Write *true* or *false* (each answer, 2 points).

1. ______ Mold can produce offspring through vegetative reproduction.
2. ______ Yeast forms spores inside spore cases.
3. _____ Fruit holds seeds.
4. _____ The body of mold is called hyphae.
5. _____ Ferns produce their own food.
6. _____ Mushrooms have spore cases.
7. _____ Pollen carries sperm.
8. _____ Ferns have pollen.
9. _____ Eggs are formed in ovaries.
10. _____ The egg moves to the sperm for fertilization.
11. _____ There are female pine cones.
12. _____ Petals dry up after fertilization occurs.
13. _____ Yeast is a fungus.
14. _____ There are male spores.
15. _____ Yeast reproduces through mitosis.

Complete this activity (each answer, 3 points).

16. Put these fern life cycle events in order.

fern matures
tiny fern grows
prothallus forms
sperm and egg develop
spore begins growth
fertilization
many spores are formed

a. ______________________________

b. ______________________________

c. ______________________________

d. ______________________________

e. ______________________________

f. ______________________________

g. ______________________________

Complete these statements (each answer, 3 points).

17. Cell division is called ______________________.
18. Sperm can be carried by ____________________ grains.
19. When sperm joins an egg, ____________________ takes place.
20. The fern's tiny, flat plant called a prothallus grows during the ____________________ stage.
21. Bees are important to flowers because they carry ______________ from one part to another.

Write the correct answers (each numbered item, 4 points).

22. Explain vegetative reproduction. ____

____

____

23. How do seeds get from one location to another? __________________________________

____

____

____

24. Name two male parts and three female parts of a flower.

a. female ____________________ d. male ____________________

b. female ____________________ e. male ____________________

c. female ____________________

Match these items (each answer, 2 points).

25. _______ petals of a flower
26. _______ spore-bearing plant
27. _______ reproductive parts of a flower
28. _______ life stages of any group of plants
29. _______ one-celled fungi
30. _______ anther, pollen, pistil, and stamens
31. _______ female cell
32. _______ life stages of any single plant
33. _______ new cell from lump on parent cell
34. _______ important to the earth's oxygen supply
35. _______ male cell

a. yeast
b. egg
c. algae
d. hyphae
e. budding
f. life cycle
g. fern
h. pistil and stamen
i. neither male nor female
j. sperm
k. seed-bearing plant
l. lifetime

Date ____________________
Score ____________________
Possible Score ______100______

Name ______________________________

Write *true* or *false* (each answer, 2 points).

1. _____ Mammals make milk to feed their young.
2. _____ Birds lay eggs with hard shells.
3. _____ A host lives in or on other animals.
4. _____ A nymph is without wings.
5. _____ Amoebas reproduce through osmosis.
6. _____ An invertebrate has a backbone.
7. _____ Pacific salmon migrate near the place they were born to reproduce.
8. _____ A mollusk has a backbone.
9. _____ Worms have scales.
10. _____ Birds are the fastest animals.

Complete these statements (each answer, 4 points).

11. A tapeworm can be a __________________ in a cat.
12. The frog is one of the common ________________________.
13. An animal in an unhatched egg is called ________________________.
14. Sperm are formed in a male animal's _____________________.
15. A paramecium moves about by means of ____________________.

Write the correct letter and answer on each line (each answer, 3 points).

16. The egg is _________________ when egg and sperm cells join together.

    a. fertilized   b. dead   c. hatched

17. Gills are needed by ____________________ in order to breathe.

    a. pupae   b. nymphs   c. tadpoles

18. There are about ______________________ kinds of insects.

    a. 5,000   b. 50,000   c. 1 million

19. A cat can get tapeworms by eating ______________________.

    a. tapeworm eggs   b. infested fleas   c. adult tapeworms

20. Fur or hair is only found on ______________________.

    a. mollusks   b. worms   c. mammals

Complete this activity (each answer, 4 points).

21. Place these life cycle events in the proper order.

    tail is lost   adult eats insects   egg hatches

    egg cell is fertilized   in tadpole form

    a. ______________________

    b ______________________

    c. ______________________

    d. ______________________

    e. ______________________

Answer these questions (each numbered item, 5 points).

22. Why is a spider not considered an insect? ______________________

    ____

    ____

23. What are mollusks? ____

____

____

24. How do fish breathe and move through water? ____________________

____

____

25. What is an amphibian? ____

____

____

26. How do mammals differ from other animals in the care of their offspring? ________

____

____

____

Date ____________________
Score ____________________
Possible Score ____100____

Science 504 Alternate Test

Name ____________________

Write *true* or *false* (each answer, 2 points).

1. _____ Producers take energy from the sun.
2. _____ Predators consume decomposers.
3. _____ Primary consumers eat mostly animals.
4. _____ Tall grass could be considered a producer.
5. _____ Rabbits are considered a secondary consumer.
6. _____ Many wolves live on the present-day prairie.
7. _____ Settlers replaced much grass with other plants.
8. _____ A plant cell has a cell wall.
9. _____ Not all living organisms need cells.
10. _____ Cells have cytoplasm.

Match these items (each answer, 2 points).

11. _______ evaporating
12. _______ photosynthesis
13. _______ More of these animals were on the early prairie.
14. _______ None of these animals were on the early prairie.
15. _______ kills and eats other animals.
16. _______ an animal that does not make the kill but eats meat
17. _______ a chemical in the chemical cycle
18. _______ protect plants from grazers
19. _______ Trees grew best in this prairie location.
20. _______ dirty air or water

a. use sun's energy
b. cows
c. producer
d. water cycle
e. predator
f. thorns
g. prey
h. creeks and streams
i. mitosis
j. eagles
k. nitrogen
l. scavenger
m. polluted
n. desert

Complete these statements (each answer, 3 points).

21. Small trees were often killed of by ____________________, but grasses usually survived.

22. Sharp leaves protected some plants from ____________________.

23. Big changes in the prairie balance of nature were made by the coming of ________________ ____________________________.

24. Low order consumers grow in numbers for a while after the killing off of ________________ ____________________________.

25. Many of the large cities seem to have a cloud of ____________________ over them.

26. Water pollution sometimes stops or slows ____________________ in plants.

Complete these items (each answer, 5 points).

27. In what ways can the prairie web of life be compared to other life systems such as the desert, pond, or forest?

____

____

____

____

28. Define stewardship as it relates to the web of life.

____

____

____

____

29. How can the hunting of animals affect the balance of nature?

____

____

____

____

Write the correct letter and answer on each line (each answer, 3 points).

30. Stewardship involves being ________________ living things.

    a. careless with   b. careful with   c. afraid of

31. Humans breathe in a. ________________ and give off b. ________________.

    a. nitrogen   b. carbon dioxide   c. oxygen

32. Plant growth can be damaged by the pollution of a. ___________ and b. ___________.

    a. air   b. noise   c. water

33. Most numerous primary consumers of the prairie were ________________________.

    a. grasshoppers   b. bison   c. deer

34. Two decomposers are bacteria and ____________________.

    a. fungi   b. rabbits   c. birds

35. Decomposers help rot ____________________.

    a. tin cans   b. rocks   c. dead animals

36. Producers, primary consumers, secondary consumers, and decomposers are all part of the ______________________.

    a. natural triangle   b. economic system   c. web of life

Date ____________________
Score ____________________
Possible Score ______100______

Name ______________________

Write *true* or *false* (each answer, 2 points).

1. _____ Work cannot be measured.
2. _____ Solar energy comes from the sun.
3. _____ Play cannot be work.
4. _____ Energy is needed for work to be done.
5. _____ Lightning is a form of electrical energy.
6. _____ Heat energy can be changed to mechanical energy.
7. _____ Nuclear energy cannot make electricity.
8. _____ All matter has some heat energy.
9. _____ Explosions cause sound energy.
10. _____ Food stores mechanical energy.

Write the correct letter and answer on each line (each answer, 3 points).

11. A burning leaf gives off ____________________ energy.
    a. heat   b. mechanical   c. potential
12. Energy cannot be ____________________.
    a. lost   b. bought   c. used
13. Solar energy does not give off ____________________.
    a. rays   b. light   c. pollution
14. Geothermal energy is produced when water touches ____________________.
    a. biomass   b. oil   c. hot rocks
15. Stored energy is ____________________ energy.
    a. kinetic   b. potential   c. old
16. Work is done when photosynthesis changes light energy to ____________________ energy.
    a. electrical   b. mechanical   c. chemical

17. Transformation of energy causes ________________.

    a. forms   b. matter   c. work

18. Mixing vinegar and baking soda creates ________________.

    a. matter   b. heat   c. oil

Match these items (each item, 2 points).

19. ______	nuclear energy	a. ability to work
20. ______	friction	b. form of electrical energy
21. ______	collision	c. made of small particles
22. ______	fuel	d. can be burned easily
23. ______	lightning	e. difficult to store
24. ______	energy	f. changes mechanical energy to heat energy
25. ______	matter	g. something moves or changes form
26. ______	solar energy	h. rubbing together
27. ______	work is done	i. using energy wisely
28. ______	stewardship	j. can cause heat pollution

Answer these questions (each answer, 5 points).

29. What is meant by *transformation of energy?* (Give two examples.) ________________

____

____

____

____

30. How does our use of energy today affect the future? ________________

____

____

____

Complete these sentences (each answer, 4 points).

31. Nuclear __________ takes place in the sun and produces great amounts of energy.
32. Burning is one way to release ________________ energy.
33. Oil stores _________________ energy.
34. Force times distance is the measure for __________________.

Answer these questions (each answer, 5 points).

35. What are two problems with the use of nuclear fission for an energy source? ___________
____
____
____
____

36. What are the two biggest uses for energy from oil? _______________________________
____
____
____
____

Date ____________________
Score ____________________
Possible Score ____100____

Science 506 Alternate Test

Name ____________________

Write *true* or *false* (each answer, 2 points).

1. ________ The life cycle was formed after the Flood.
2. ________ The Flood formed oil according to the Bible record.
3. ________ Cities were built before the Flood.
4. ________ Animals were expected to reproduce after the Flood.
5. ________ The Bible warns that God may use a flood to destroy the world again.
6. ________ Some animals became extinct.
7. ________ Mammoth remains were found in permafrost.
8. ________ Some plants grew very large before the Flood.
9. ________ Glaciers are present on earth today.
10. ________ Huge fossil deposits have been discovered.

Match these items (each answer, 3 points).

11. _____ People had long lives before the Flood.
12. _____ Glaciers came to the earth.
13. _____ Alien beings from another planet lived on earth.
14. _____ Fish fossils were found in oil deposits.
15. _____ Petrified wood was once living trees.
16. _____ Adam's remains have been found.
17. _____ Rock layers give clues to the Flood story.
18. _____ God was upset with man's wickedness.
19. _____ There were billions of people on earth when the Flood came.

a. Physical record
b. Bible record
c. Neither record

Complete this activity (each answer, 3 points).

20. Write these events in the proper order:
    glaciers formed
    much water covered the earth
    seasons affected plant growth
    present water cycle formed
    worldwide mild climate

    a. ______________________________
    b. ______________________________
    c. ______________________________
    d. ______________________________
    e. ______________________________

21. God told Noah to build a(n) ____________________.

22. Much physical evidence ________________ with the Bible accounts of the Flood.

23. Much ________________ of life took place during the Flood.

24. No ____________________ account of the Flood was made by Noah or others who lived then.

25. The ___________ cycle as we know it came about after the Flood.

26. Some complete mammoth remains were found in ____________________.

27. An example of a shifting landmass is the American ____________________.

Answer these questions (each answer, 4 points).

28. Why did God want to make a new beginning with Noah? ____________________________
____
____
____

29. How does physical evidence show that a flood covered the earth? ____________________

____

____

____

____

____

Write the correct letter and answer on each line (each answer, 3 points).

30. The hardened remains of a plant or animal are ____________________.

    a. thistles   b. fossils   c. glaciers

31. Oil deposits were formed from ____________________.

    a. animals   b. fossils   c. rocks

32. Earthquakes and ____________________ helped make changes in the earth.

    a. mitosis   b. volcanoes   c. canyons

Date ____________________
Score ____________________
Possible Score ____100____

Name ____________________

Match these items (each item, 2 points).

1. _____ "intelligent design"
2. _____ print fossils
3. _____ original-remains
4. _____ petrified fossils
5. _____ carbonized fossils
6. _____ tar pits
7. _____ fossil discoveries
8. _____ mammoth fossils
9. _____ method
10. _____ Creation scientist

a. minerals replace some or all of original materials
b. guided Creation and development of all things
c. believes the biblical record
d. broken bits
e. hair and scales
f. carbon remains
g. the way to do something
h. insects in amber
i. happen often
j. located in Los Angeles
k. found in Siberia
l. cast and mold

Write *true* or *false* (each answer, 2 points).

11. ____________ Fossil bones have not been found in caves.
12. ____________ Coal is not a fossil fuel.
13. ____________ A famous fossil find is located in the Gobi Desert.
14. ____________ A new building site may be a good place to look for fossils.
15. ____________ Limestone is not a type of sediment.
16. ____________ Dinosaur eggs have been found.
17. ____________ Fossil type is one way to identify fossils.
18. ____________ No fossils have been found of extinct animals.
19. ____________ Many fern fossils have been found.
20. ____________ Dating of fossils by age is sometimes used to support theories of evolution.

Write the correct letter and answer on each line (each answer, 3 points).

21. ____________________ fossils are usually found in coal.

    a. No b. Carbonized c. Print

22. When looking for fossils, ____________________ is important.

    a. safety b. having a cell phone c. a car

23. Fossils from ____________________ are common.

    a. donkeys b. flowers c. seed-bearing plants

24. Clues give information so we can make ____________________.

    a. conclusions b. cast molds c. money

Complete these lists (each answer, 4 points).

25. List three ways to identify fossils.

    a. ____________________

    b. ____________________

    c. ____________________

26. List two types of print fossils.

    a. ____________________

    b. ____________________

27. List two types of rock more likely to hold fossils.

    a. ____________________

    b. ____________________

Answer these questions (each answer, 5 points).

28. How can fossil hunters practice good stewardship? ______________________________
____
____
____

29. What can reconstructed skeletons or models of fossils tell us? ____________________
____
____
____

30. What is a fossil? ____
____
____

31. Where have fossils been found? (Do not list specific locations). ____________________
____
____

Date ______________________
Score ______________________
Possible Score ____**100**____

Name ______________________

Match these items (each answer, 2 points).

1. _____ flows from volcanoes
2. _____ center of earth
3. _____ just below the crust
4. _____ melted rock inside the earth
5. _____ granite is here
6. _____ crust materials
7. _____ can break rocks open
8. _____ way of identifying rocks
9. _____ kind or rock formed from magma
10. _____ can cause erosion

a. mantle
b. lava
c. wind storms
d. igneous
e. metamorphic
f. growing plants
g. color
h. outer core
i. rocks, soil
j. under earth's land areas
k. magma
l. dense ball

Write *true* or *false* (each answer, 2 points).

11. _____ The earth is slightly flattened at the poles
12. _____ The earth's size can be determined by using mathematical methods.
13. _____ Oceans are not deep when compared to the total earth.
14. _____ Surface features are in constant change.
15. _____ The earth's size is getting much smaller.
16. _____ Magma seldom flows.
17. _____ Some rocks have less luster than other rocks.
18. _____ Sedimentary rocks were formed from layers of sand and soil.
19. _____ Identifying rocks can help us understand the earth's history.
20. _____ A conglomerate rock can have several pebbles in it.

ite the correct letter and answer on each line (each answer, 3 points).

21. The earth is shaped like a ______________________.

    a. sphere b. spear c. shear

22. Columbus thought the earth was ____________ than it really is.

    a. smaller b. older c. colder

23. The form of most _____________ is crystal.

    a. chemicals b. minerals c. faults

24. Earthquakes happen when rocks at ________________ lines shift.

    a. power b. fold c. fault

25. Movement in the earth's surface can be caused by __________________ from below.

    a. pressure b. explosions c. air current

26. Changes of ___________________ can cause rocks to chip apart.

    a. living conditions b. temperature c. streak

27. Magma flowing between rock layers has helped ______________ landforms.

    a. shake b. shape c. shale

28. Earthquakes have caused ___________________ to form.

    a. lakes b. life c. gravity

omplete these descriptions (each description, 5 points).

29. Describe how forces can build up a landform while other forces wear it away.

    ____

    ____

    ____

30. Describe how folding has affected living conditions for sea life.

____

____

____

31. Describe how igneous rocks are formed.

____

____

____

32. Describe what can happen during an earthquake.

____

____

____

Complete this list (each answer, 4 points).

33. List four physical tests used to identify rocks.

a. ____________________________________

b. ____________________________________

c. ____________________________________

d. ____________________________________

Date ____________________
Score ____________________
Possible Score ______100______

Name ____________________

Match these items (each item, 2 points).

1. _____ water vapor
2. _____ the most heat causes this
3. _____ takes the shape of its container
4. _____ molecules grouped closely together
5. _____ molecules moving fastest
6. _____ molecules locked together
7. _____ often matter cannot be seen
8. _____ result of taking heat away from gas state
9. _____ clouds
10. _____ snow

a. solid state
b. liquid state
c. gas state

Write *true* or *false* (each answer, 2 points).

11. __________ An object has the same mass on the moon as on the earth.
12. __________ All matter has the same special properties.
13. __________ Two items can take up the same space.
14. __________ To find an item's special properties, tests must be made.
15. __________ A special property of matter is brittleness.
16. __________ Rust is a physical change.
17. __________ Atoms are made up of compounds.
18. __________ The head of a comet has a nucleus.
19. __________ Formation and decay help cycle matter.

List these items (each answer, 4 points).

20. List the parts a comet may have.

    a. ______________________________

    b. ______________________________

    c. ______________________________

    d. ______________________________

Write the correct letter and answer on each line (each answer, 3 points).

21. When matter that is still stays still, the ____________ property causes it not to move.

    a. stretch b. chemical c. inertia

22. The amount of space taken up by something is called its ________________.

    a. mass b. volume c. distance

23. The ____________________ of the earth is one reason why seasons change.

    a. properties b. slanting c. axis

24. The conservation of matter means that matter is not being ________________ now.

    a. changed b. created or lost c. lost

25. The Bible gives an example of the balance in God's creation of the ____________ cycle.

    a. water b. mineral c. carbon

26. The Bible uses __________________ to describe the earth's chemical nature.

    a. hydrogen b. dust c. prediction

27. Clouds are formed when air cools to ____________________.

    a. crystals b. evaporation c. the dew point

Answer these questions (each answer, 5 points).

28. What changes take place when something burns?

____

____

____

29. What is shown about the Creation when the Bible says that God provided plenty of those things needed to support life?

____

____

____

30. Why are changing seasons seen as a cycle of nature?

____

____

____

31. How are chemical changes different from physical changes?

____

____

____

32. Why are formation and decay considered a cycle of nature?

____

____

____

Date ____________________
Score ____________________
Possible Score _____**100**_____

Science 510 Alternate Test

Name ____________________

Write *true* or *false* (each answer, 2 points).

1. __________ A larva and pupa are the same thing.
2. __________ A spore is an important part of a fungus life cycle.
3. __________ Some vertebrates provide their young with milk.
4. __________ A decomposer is really a parasite.
5. __________ A seed-bearing plant may have cones instead of flowers.
6. __________ Plant cells have cell walls but no cell membranes.
7. __________ Mitosis is important to cell reproduction.
8. __________ Each cell performs work.
9. __________ The chemical cycle is part of the balance of nature.
10. __________ Matter is not being created at present.

Match these items (each answer, 2 points).

11. _____ chlorophyll
12. _____ decomposer
13. _____ primary consumer
14. _____ secondary consumer
15. _____ producer
16. _____ print fossil
17. _____ carbonized fossil
18. _____ weathering
19. _____ flooding
20. _____ state of matter

a. grass
b. liquid
c. red
d. bacteria
e. carbon remain
f. complete mammoth
g. wolf
h. in chloroplasts
i. wind and rain
j. insect
k. cast or shell
l. erosion

Write the correct letter and answer on each line (each answer, 3 points).

21. The earth has more ____________ plants than any other type.

    a. seed-bearing   b. spore-bearing   c. one-celled

22. Animal cells in tissue vary in ________________ and size.

    a. color   b. makeup   c. shape

23. A force from within the earth that changes the surface is ________________.

    a. folding   b. a glacier   c. erosion

24. The details of fossil formation are ________________ the Bible.

    a. inferred in   b. left out of   c. given in

25. Burning causes matter to change ________________.

    a. chemically   b. physically   c. partly

26. A special property of matter is ________________.

    a. mass   b. ability to take up space   c. solubility

27. List five types of energy (3 points).

    a. ____

    b. ____

    c. ____

    d. ____

    e. ____

28. List four types of landforms.

a. ____________________

b. ____________________

c. ____________________

d. ____________________

Answer these questions (each answer, 5 points).

29. How could the Flood have formed sedimentary layers?

____

____

30. What is energy transformation?

____

____

____

31. How is mitosis important to plants and animals?

____

____

____

Date ____________________
Score ____________________
Possible Score ____100____

LIFEPAC

ANSWER KEYS

SECTION ONE

1.1 cells

1.2 cork

1.3 basic unit

1.4 unicellular

1.5 multicellular

1.6 b

1.7 f

1.8 a

1.9 d

1.10 g

1.11 c

1.12 Typical 3-part cell

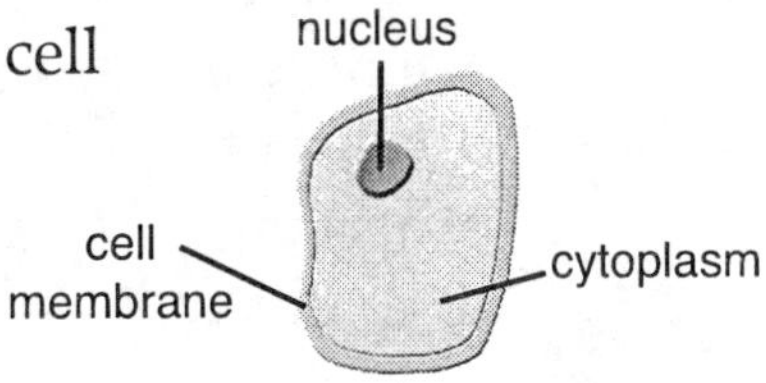

nucleus

cell membrane

cytoplasm

1.13 true

1.14 false

1.15 true

1.16 true

1.17 false

1.18 Microscopes help us to view cells. (Two types of microscopes are optical microscopes and electron microscopes.) It is also helpful to use dyes to view cells.

1.19 Study the drawing on page 9 of the LIFEPAC

1.20 The student's additional observations should be noted.

SECTION TWO

2.1 cell membrane

2.2 a. phospholipid
b. proteins

2.3 organelles

2.4 nuclear membrane

2.5 Either answer is acceptable: chromatin or chromosomes

2.6 DNA

2.7 chromosomes

2.8 nucleolus

2.9 Any order:
a. those that produce proteins
b. those that produce energy
c. specialty organelles

2.10 DNA and genes contain the molecular information to make the cells and groups of cells within a living thing what they are to be. The DNA is what makes the offspring of a living thing like the parent.

2.11 false

2.12 true

2.13 true

2.14 true

2.15 false

2.16 true

2.17 Onion bulb cell

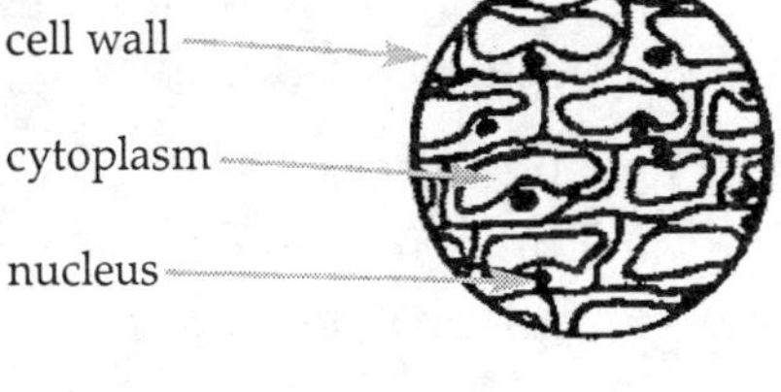

cell wall

cytoplasm

nucleus

2.18 Any other information the student found interesting is acceptable.

2.19 Pond water first slide:
Compare with drawings of the unicelluar animals in the LIFEPAC.

2.20 Pond water second slide
Compare with drawings of the unicelluar animals in the LIFEPAC.

2.21 Pond water third slide
Compare with drawings of the unicelluar animals in the LIFEPAC.

2.22 For a., b. and c., compare with drawings of the unicelluar animals in the LIFEPAC. Some answers may be "unknown."

2.23 Any other information the student found interesting is acceptable.

2.24 a. membrane
b. protein
c. chromosomes
d. chlorophyll
e. unicellular
f. phospholipid
g. nucleolus
h. cellulose
i. photosynthesis
j. protozoa

2.25 Cheek cells: refer to epithelial cells on page 25.

2.26 Any other information the student found interesting is acceptable.

2.27 They should appear to be alike in structure. Their composition is alike.

2.28 There should not be basic differences. Perhaps size differences would appear because of the drawing size.

2.29 The functions of the cells are similar. The functions of the cells are to cover and protect.

2.30 Blood cells: refer to red blood cells on page 24.

2.31 Any other information the student found interesting is acceptable.

2.32 unicellular

2.33 multicellular

2.34 White

2.35 Red

2.36 Nerve

2.37 Epithelial

2.38 Muscle

2.39 Any order:
a. epidermal
b. connective
c. storage
d. support

2.40 Any order:
a. epithelial
b. muscular
c. nervous
d. connective

2.41 A tissue is a group of cells in a multicellular plant or animal that is similar in structure and performs similar functions.

2.42 Some of the functions are similar — epithelial tissue covering in both. Connective tissue helps move needed nutrients in both plants and animals.

2.43 Some of the following should be covered: The nerve cells in the nervous tissue are close together and can send signals, or impulses, from one to another very quickly. Nervous tissue is located all through the body. It forms the communication network to and from the brain. Sensory nervous tissue is responsible for sending information to the brain. This sensory information comes from nerve cells and nerve tissues located in the eyes, ear, nose, mouth, and skin. The brain then receives and processes these messages and information. Then, information is sent out from the brain through motor nervous tissues in order to move muscles, activate certain glands, or perform other body functions.

2.44 Some of these thoughts could be shared: The body is made up of complex cells and tissues. They are wonderful in the way they are structured and function. Even the tiny cells are very complex. God has made us wonderfully.

SECTION THREE

3.1 Plants take in carbon dioxide through their leaves and water from their roots. In photosynthesis, the chlorophyll in the plant receives the energy needed from the sun to cause a chemical reaction with the carbon dioxide and water. As a result, oxygen gas is produced, along with sugars and other materials that the plant can use as food. The oxygen produced by photosynthesis is given off through the leaves.

3.2 No.

3.3 Respiration is the opposite of photosynthesis. Oxygen is used up, and water and carbon dioxide are given off.

3.4 c. oxygen and sugars

3.5 b. both plants and animals

3.6 d. red blood cells

3.7 c. energy

3.8 a. each other

3.9 Food is brought into the body through eating and the body's digestive system. Oxygen is brought into the body through breathing. Respiration occurs when the food is combined with the oxygen in the body, giving off energy the body needs to perform life and work.

3.10 Plants and animals depend on each other to carry on life. Plants must have adequate carbon dioxide given off by animals and human beings in order for the process of photosynthesis to take place. In turn, animals and human beings rely on plants for the oxygen and much of the food they receive. God has arranged this important cycle of energy in the world.

3.11 false

3.12 true

3.13 true

3.14 false

3.15 true

3.16 true

3.17 two

3.18 mitosis

3.19 cell division

3.20 red blood

3.21 Nerve

3.22 Adult check

SECTION ONE

1.1 God

1.2 kingdoms

1.3 kingdoms; other answers in any order:
a. animals
b. plants
c. fungi
d. protists
d. monerans

1.4 Some of the basic characteristics include physical structure and make-up, the means of obtaining food, and the means of reproduction.

1.5 Fungi do not contain chlorophyll as green plants do. Fungi must also obtain their food from outside sources, while plants produce their own food.

1.6 See Table I in Section One for several examples of organisms in each kingdom.

1.7 Adult check

1.8 true

1.9 false

1.10 false

1.11 true

1.12 true

1.13 true

1.14 false

1.15 Any order:
a. epidermal
b. connective
c. storage
d. support

1.16 Any order:
a. roots
b. leaves
c. stems
d. flowers

1.17 a. beginning
b. growth
c. adulthood
d. death — or end

SECTION TWO

2.1 seed-bearing plants

2.2 flowering

2.3 gymnosperms

2.4 fertilization

2.5 pollen

2.6 In mitosis, one cell splits apart to form two new cells. Mitosis is the way most cells reproduce.

2.7 Use Science LIFEPAC 501, Section III. One of the most common ways that cells reproduce is called **mitosis**. It occurs in eukaryote cells only; that is, cells that have a cell membrane, cyptoplasm and a nucleus. Mitosis is the process of one cell splitting apart to form two new cells. This "splitting apart" of the cells in known as *cell division*. Mitosis brings about cell division and two new cells from one original cell.

Mitosis starts when the chromatin within the cell begins to rearrange and condense into orderly strands called *chromosomes*. The chromosomes then move into pairs. After that, the chromosome pairs begin to pull apart from each other. Eventually, the chromosome pairs split apart. When they split apart, cell division occurs. Then there are two new cells instead of the original single cell. The two new cells are smaller than the original cell. However, because the DNA of the two new cells is the same as the original cell, the two new cells will be like the original cell, only smaller. In time, the two new cells will grow larger in size, until they also begin the process of mitosis and split into new cells.

Mitosis is the way that most one-celled living things reproduce. New individual cells are formed whenever an individual cell divides through mitosis. The new cells appear to be just like the former cells, although they may be a bit smaller at first.

Mitosis also happens when cells are worn out, damaged, or need to be replaced. Skin cells and red blood cells are examples of cells that can replace themselves. Nerve cells, however, cannot be replaced when they are damaged. Some animals can replace lost body parts through mitosis!

2.8 a

2.9 a

2.10 b

2.11 b

2.12 c

2.13 c

2.14 d

2.15 Adult check

2.16 Adult check – Use the diagram of the cross-section of a flower of page 16 of the LIFEPAC to help you.

2.17 true

2.18 false

2.19 true

2.20 true

2.21 false

2.22 true

2.23 Jesus meant that, just as God took care of the lilies of the field, He would take care of us. He compared the splendor of the lilies (flowers) to the splendor of Solomon's clothing.

2.24 Record the similarities of the flowers

2.25 Record the differences of the flowers.

2.26 a. stigma
b. ovary
c. style

2.27 a. filament
b. anther

2.28 Pollen is formed in the anther.

2.29 Pollen could get to the stigma by wind blowing it or by bees or insects carrying it.

2.30 The stigma holds the pollen for a time. The pollen forms a tube leading down to the pistil to the ovary.

2.31 You should have counted the seeds in your sample fruit.

2.32 drawing of seeds

2.33 measurements of the different seeds

2.34 description of how seeds could travel

2.35 Adult check

2.36 Adult check

2.37 b

2.38 a

2.39 d

2.40 c

2.41 f

2.42 Any order:
a. carry the seeds and protect them
b. help seeds move from place to place
c. food source for animals and people
d.provide beauty and enjoyment in nature

2.43 The farmer cuts the white potato into many parts, making sure each part has at least one "eye" (bud). Each piece of potato will grow from this part into an entirely new potato plant.

2.44 Dandelions are caught by the wind and move from place to place. Some fruits are eaten by birds and animals, digested, and the seeds are deposited elsewhere through droppings. Squirrels move nuts from place to place. Some seeds are caught in animals' fur and are moved to another place. Finally, people eat fruit and throw the seeds elsewhere.

2.45 a. female
b. male

2.46 sticky fluid

2.47 scales

2.48 pollen

2.49 a. sperm cell
b. egg cell

2.50 false

2.51 true

2.52 false

2.53 true

2.54 true

2.55 To protect seeds. To make seeds. To move seeds.

2.56 Cones do not form fruit. Cones don't have pistils, stamens, anthers, or stigmas.

2.57 Seed poster

SECTION THREE

3.1 Any five:

mosses	fungi
ferns	algae
toadstools	bacteria
mushrooms	protozoans
molds	

3.2 a. beginning
b. growth
c. adult
d. old age and death

3.3 Growth of the organism happens by mitosis. Moisture helps the cells divide. However, the spores must also be attached to something that will provide food for them. They cannot produce their own food. When the spore has enough moisture and food provided for it, mitosis continues and the spore grows into an adult organism. The growth stage of the organism can be either in two parts or one part, depending upon the type of parent organism.

3.4 b. a spore

3.5 a. spore cases

3.6 b. mitosis

3.7 c. the prothallus

3.8 c. swimming through water

3.9 false

3.10 true

3.11 true

3.12 false

3.13 false

3.14 false

3.15 southern United States

3.16 damp and wet

3.17 yes

3.18 tangled masses of tiny threads that make up fungus plants

3.19 from other plants or dead organisms

3.20 in spore casings at the end of stalks

3.21 Ferns make their own food. Ferns have an added step in the growth stage — the prothallus. Fungi do not need fertilization.

3.22 Daily Observation Chart

3.23 your explanation of the life cycle

3.24 Possible labels might be the stalks and the spore cases.

3.25 Adult check

3.26 Adult check

SECTION FOUR

4.1 chlorophyll

4.2 mitosis

4.3 a. food
b. oxygen

4.4 smaller

4.5 would weigh more

4.6 A lump forms on the parent cell. Part of the nucleus moves to the cell. The cell membrane will soon pinch away from the bud. Two yeast cells are produced.

4.7 Yeasts are oval and not green. Algae are reproduced through mitosis. Growth and reproduction happen quickly. Many algae are formed.

4.8 Each cell is able to do all the functions needed to keep alive. If any yeast cell would separate from the others, it could live without the other cells.

SECTION ONE

1.1 five
1.2 zoology
1.3 1 1/2 million
1.4 Cold-blooded
1.5 omnivores
1.6 invertebrates
1.7 move around
1.8 The life cycle of a living thing refers to the different life stages that it goes through during a normal lifetime. Typically, these are beginning, growth, adulthood, old age, and death.
1.9 It does not have a backbone.
1.10 c
1.11 d
1.12 b
1.13 e
1.14 a
1.15 true
1.16 false
1.17 true
1.18 true
1.19 true
1.20 The adult stage of an invertebrate is reached when it grows to full size and is able to reproduce. It looks very much like its parents. Its form will change very little during the adult stage. It can begin to reproduce. Some invertebrates will reproduce many times during their adult stage.
1.21 Most invertebrates do not reach old age. They die at an earlier stage in the life cycle. The harsh conditions of changing weather can cause the death of invertebrates. Larger animals feed on invertebrates. Other reasons may cause an early death. Yet, enough of these animals survive so that life can continue as they grow to adulthood and reproduce. They usually lay many more fertilized eggs than needed so that life can continue for each kind of invertebrate.
1.22 c. 30,000
1.23 c. a "false foot"
1.24 b. tiny hairs
1.25 b. cell membrane
1.26 a. oxygen
1.27 c. mitosis
1.28 Adult check
1.29 Adult check
1.30 a
1.31 b
1.32 a
1.33 a
1.34 c
1.35 c
1.36 1 million
1.37 testes
1.38 ovaries
1.39 larva
1.40 Adult check
1.41 Adult check
1.42 Adult check
1.43 Adult check
1.44 Adult check
1.45 Spiders do not have six legs. They have only two body parts. There are no wings.

1.46 Beginning — males mate with females and the eggs are fertilized. (Males do not guard the female or eggs.)

Growth — eggs hatch. As baby spiders grow, their skins get too tight — they shed them.

Adult — full-grown spiders can reproduce. Life span is short. Females live longer.

1.47 It is similar to the insects that have a simple cycle. There are no larvae, pupae, or nymphs. They lay eggs and grow through mitosis.

1.48 Adult check

1.49 Worms are animals that have soft, slender bodies and no backbone or legs. Worms have no outside covers or bones to give them protection. Since worms have no protective structures, they live in places that are safer for them. Most of their lives are spent under the ground, in water, or inside other animals.

1.50 There are big differences in the life cycles of real worms and the larvae of insects. Larvae will change into adult insects sometime during the life cycle. The adult insects no longer look like worms. Worms will stay worms all their lives. The adult worms can reproduce. Insect larvae cannot reproduce.

1.51 Parasites are animals that live on or in other animals. They get their food from the hosts.

1.52 The hosts for the tapeworm larvae are usually fleas. The host fleas are infested with the tapeworm larvae. When a cat cleans itself, it swallows the fleas. The tapeworm larvae on the fleas change into tiny tapeworms. They then live and grow as parasites in the intestines of the cat. The tapeworms produce eggs and fertilize them with sperm. The fertilized eggs are carried out of the cat's body as waste.

Fleas feed on animal waste. If tapeworm eggs are in the waste, the eggs are brought into the fleas where they hatch into larvae. When the cat swallows the larvae-infested fleas, the tapeworm life cycle continues.

1.53 A mollusk is a soft-bodied invertebrate animal that has no bones. Most species of mollusks grow hard shells to protect themselves.

1.54 They both lay eggs. Both include species that have larvae.

1.55 Mollusks with larvae do not have a pupae form. Adult mollusks grow shells. Insects do not.

1.56 Some mollusks become parasites when they are larvae.

1.57 Across:
- b. mollusk
- e. parasite
- i. testes
- k. cycle
- m. pupa

Down:
- a. worm
- c. larva
- d. cell
- f. amoeba
- g. insect
- h. mate
- j. squid
- l. egg

SECTION TWO

2.1 backbones

2.2 body

2.3 embryo

2.4 can reproduce

2.5 reproduce

2.6 true

2.7 true

2.8 true

2.9 false

2.10 true

2.11 false

2.12 true

2.13 true

2.14 Adult check

2.15 It is an animal that spends part of its life as a water animal and part of its life as a land animal.

2.16 about 4,000 kinds

2.17 A frog is like a fish when it is in the tadpole form.

2.18 The male gives out a mating call. The female finds the male.

2.19 Metamorphosis is the process by which a tadpole changes into a frog. During metamorphosis, its lungs develop, legs grow, the tail shrinks and disappears, and the mouth gets much larger. Eventually the tadpole becomes a frog and it leaves the water to live on land.

2.20 b. amphibian

2.21 c. reptile

2.22 a. inside the egg

2.23 b. lizards and snakes

2.24 The birds move from place to place seeking warmer climate. Usually, it involves long distances during different seasons of the year.

2.25 The "lifeline" could be in the form of a chart or graph and should include the following information. In the spring, the egg is fertilized. Soon the female lays the egg. In two weeks the egg is hatched. The robin grows enough to survive. The robin is pushed from the nest. The robin stays near the nest the first summer. At the end of the summer the robin migrates south. The next spring the robin returns north to reproduce.

2.26 Adult check

2.27 Adult check

2.28 Adult check

2.29 Adult check

2.30 It needs to have a way for the air to reach the embryo. The embryo needs oxygen for energy to grow.

2.31 It protects the egg shell. It allows needed supplies in and allows waste (carbon dioxide) out.

2.32 a. It protects the yolk and embryo.
b. It supplies food for the embryo.
c. It is the part that will grow through mitosis into a chick.

2.33 Adult check

2.34 Adult check

2.35 Adult check

2.36 Adult check

2.37 Adult check

2.38 Adult check

2.39 false

2.40 false

2.41 true

2.42 true

2.43 true

2.44 Any order:

a. Only mammals nurse their babies on mother's milk.
b. Only mammals have hair.
c. Mammals are warm-blooded.
d. Mammals have a larger, more well-developed brain.
e. Most mammals give their offspring training and protection.

2.45 Adult check

2.46 Adult check

SECTION ONE

1.1 water, air, soil, rocks, minerals, and chemicals.

1.2 A balance of nature occurs when the life needs of all the living things in an area of the earth are met.

1.3 The physical environment consists of nonliving things like water, the air, the soil, and the weather. The biological environment consists of the living things in an area of the earth.

1.4 God has created many creatures throughout the earth. He supplies all they need for life. He provides for all creatures in a way to allow a balance of nature.

1.5 water

1.6 water

1.7 evaporates

1.8 water

1.9 drought

1.10 balance

1.11 Water evaporates from the oceans, lakes, rivers, and streams to the air. The water vapor in the air cools and forms clouds. Eventually, it produces a form of precipitation like rain, snow, or dew and falls to the earth. The rain then collects in streams, rivers, lakes, and oceans. The water cycle starts again.

1.12 Adult check

1.13 true

1.14 true

1.15 false

1.16 true

1.17 true

1.18 c. all of these

1.19 c. roots

1.20 a. proteins

1.21 b. bacteria

1.22 a. chemical cycle

1.23 Adam

1.24 death

1.25 Their bodies will return to dust.

1.26 bacteria

1.27 Adult check

1.28 b. ecosystem

1.29 a. producer

1.30 c. birds

1.31 a. digest green plants

1.32 b. a pine tree

1.33 Producer: lettuce

Primary Consumer:
chicken, monkey, cow, mouse, silverfish

Secondary Consumer:
tiger, bull snake, salmon

Decomposer: mushroom

1.34 Adult check

1.35 Adult check

1.36 Most ecosystems have a great variety of producers, consumers and decomposers. These form an overlapping network of food chains called a food web.

1.37 The web of life depends on the food chain. Animals cannot produce their own food or energy. They must eat plants to get the energy. Some animals cannot digest plants. Those animals must eat other animals. Animals would all die without plants.

1.38 Water is needed for cell life. The plants cannot get water from the water cycle when they are indoors.

1.39 The terrarium will not have enough food in it. Some of the animals may be meat eaters.

1.40 The balance between plants and animals may not exist. More oxygen or carbon dioxide may be needed.

1.41 The plants need sunlight for photosynthesis to work.

1.42 Adult check

SECTION TWO

2.1 U.S.: Idaho, Montana, Wyoming, Minnesota, Wisconsin, Michigan, North Dakota, South Dakota, Iowa, Indiana, Illinois, Kansas, Nebraska, Missouri, Colorado, Utah, Arizona, New Mexico, Oklahoma, Texas

Canada: Alberta, Saskatchewan, Manitoba, Ontario, Quebec, Newfoundland

2.2 Rainfall was moderate. A struggle often existed between grasslands and forests. When a moderate rainfall area had wetter years, many trees began growing. When there were drier years, fires often occurred and killed off the small trees. However, the fires did not kill most grasses. Therefore, prairies with grasslands won out in the open areas. Where the areas were more moist and protected, forests grew.

Rainfall averaged between 25 cm per year and 100 cm per year in the prairie regions. The eastern parts of the prairie usually received more rainfall than the western parts of the prairie. The difference in rainfall is one reason why the food chain varied somewhat from east to west in the prairie ecosystem.

2.3 Both plants and animals could survive in great numbers on the prairie. The animals gave off carbon dioxide. The plants gave off oxygen. Living things had a plentiful supply of both of these life needs. The carbon cycle worked well.

2.4 The decomposers were also busy on the prairie. As plants and animals died, the chemicals in their bodies were returned to the soil. The decomposers, like bacteria, helped this process. The prairie land was a rich source of these chemicals and minerals. Plants such as clover provided much nitrogen to the soil. The chemical cycle worked well on the prairie.

2.5 rain

2.6 root

2.7 a. woody
b. sharp
c. thorns

2.8 rainfall

2.9 a. food
b. oxygen

2.10 true

2.11 true

2.12 false

2.13 true

2.14 false

2.15 false

2.16 Adult check

2.17 Adult check

2.18 Adult check

2.19 Adult check

2.20 Adult check

2.21 a. wet years
b. dry years
c. seasons or predators

2.22 weak, sickly, young

2.23 Spring and summer were reproductive seasons. Plant food was plentiful.

2.24 dies

2.25 very important to God

2.26 a. everything
b. us

2.27 Adult check

2.28 The soil is rich in minerals. There is a good amount of rainfall.

2.29 They could be killed off completely by hunters or farmers.

2.30 Adult check

SECTION THREE

3.1 When people move into an area, loss of plant and animal life usually occurs. Land is taken over for buildings. Farmers clear land for their crops. As cities begin to develop, large areas of land are covered with concrete. Hunters kill animals for food. They also kill dangerous predators.

People have many uses for the wood of trees. Sometimes, trees stand in the way of roads, farms, or buildings. People remove the trees, and this action causes a change for some animals.

3.2 It is harmful when hunters shoot all kinds of animals and too many of them. When that happens, some predators can be killed off. Some animal groups like bison can be destroyed.

3.3 Humans can be helpful in keeping a balance of nature, especially if other predators are scarce, and if the hunters know what animals to shoot and how many they should take.

3.4 photosynthesis.

3.5 oxygen

3.6 a. air
b. water

3.7 chemicals

3.8 prevents air pollution from cars

3.9 Air may stay cleaner. Litter would not get into water. Litter would not damage plants.

3.10 People would be showing care so that life can continue.

3.11 Too many animals would not be killed. Some animal groups would no longer be destroyed.

3.12 Water for plant and animal needs could be saved.

3.13 God's help would be sought in solving the pollution problem.

3.14 Adult check

3.15 Adult check

3.16 Adult check

SECTION ONE

1.1 energy

1.2 a. ability
b. work

1.3 potential

1.4 kinetic

1.5 potential energy

1.6 kinetic energy

1.7 a. -e., any order:
a. wood
b. food
c. gas
d. solar
e. nuclear
f. -k., any order:
f. sun
g. coal
h. wind
i. oil
j. water
k. chemical

1.8 d

1.9 c

1.10 b

1.11 e

1.12 f

1.13 c

1.14 a

1.15 a

1.16 a. Yes
b. Your eyelids moved. Movement involves work.

1.17 a. Yes
b. The book was moved. Movement involves work.

1.18 a. No
b. There was no movement

1.19 a. Yes
b. Your arm, hand, and the pencil moved. Movement involves work.

1.20 a. Yes
b. The paper's form and shape were changed. When the shape or form of matter change, work is done.

1.21 When a checker is moved, work is being done. The movement of the checker is work.

1.22 Work is being done when the feather is moved. This movement involves work.

1.23 The man is moving. This involves work. The child is also moving and this involves work, too. If sounds are made, work is done.

1.24 500 foot-pounds

1.25 50 newton-meters

1.26 None. Since the rock did not move any distance, no work is done.

1.27 The work of God is for people to have faith in Jesus.

1.28 God's work is important because He gives life to the world through Jesus. Anyone who believes in Jesus can live with God. This is His work.

1.29 God's work is not a movement or changing from one energy form to another. It is a change of life.

SECTION TWO

2.1 Examples:
There was heat in the drawer. The air in the drawer had heat. The air was not heated as much as the sunlight.

2.2 Answer depends on the conditions of the experiment.

2.3 Example:
More heat is absorbed from the direct rays of the sunlight.

2.4 Example:
The thermometer reading was not as high as in direct sunlight. But it was higher than when the reading was taken inside the drawer.

2.5 Adult check

2.6 Examples:
It was very warm. It was hot.

2.7 Examples:
It was cooler than the bent one. It did not get hot.

2.8 Example:
It would probably get warm or hot like the other paper clip.

2.9 Movement of the hands to bend the object was the source of mechanical energy.

2.10 true

2.11 true

2.12 true

2.13 false

2.14 true

2.15 A bubbling action should be seen when they are mixed together.

2.16 Sometimes the mixing of chemicals causes heat energy to be formed.

2.17 mixing chemicals

2.18 c. heat

2.19 b. friction

2.20 b. friction

2.21 b. chemical

2.22 a. reflects off

2.23 b. chemical

2.24 true

2.25 true

2.26 true

2.27 false

2.28 true

2.29 true

2.30 Examples:
United States, Russia

2.31 Examples:
United States, Iran, Saudi Arabia, Russia, Argentina

2.32 Example:
United States
Russia

2.33 Example:
They can develop their own industries without depending on other countries. Transportation will be less costly for countries with these resources. Other countries will pay for oil and coal.

2.34 heat energy

2.35 chemicals

2.36 mixed

2.37 photosynthesis

2.38 Either order:
a. oil
b. coal

2.39 a. heat

2.40 c. work

2.41 a. burning

2.42 a. batteries

SECTION THREE

3.1 false

3.2 true

3.3 false

3.4 false

3.5 true

3.6 true

3.7 false

3.8 true

3.9 Examples:
The skill of taking care of something for someone else. God wants us to be good stewards of the earth, and we have stewardship of the earth.

3.10 God wanted humans to take care of all the living things on earth.

3.11 Adam was supposed to till the ground.

3.12 It warms up as it stays in the sun.

3.13 Example:
The heat leaves the water and transfers to the air and other matter around it.

3.14 Examples:
You could wrap the jar in a lot of papers.
You put it in a thermos.
You held your hands on the jar.

3.15 a. solar energy

3.16 b. cost

3.17 c. interrupt

3.18 a. other forms of energy

3.19 It uses less fuel and has much less pollution.

3.20 Uranium and other radioactive elements are limited and could be used up. Harmful radiation. Possibility of nuclear accidents. Hot waste water may damage the environment. Radioactive wastes remain dangerous for long periods of time.

3.21 In fission, the atomic nucleus of an element, such as uranium, is split. The splitting of the nucleus of an atom produces a large amount of energy. Fusion involves the combining of atomic nuclei to form heavier nuclei. In this process, enormous amounts of energy are released.

3.22 Fuel source is almost unlimited in the world's oceans. Fusion devices are safer than those used in fission. Fusion does not create a waste disposal problem.

3.23 true

3.24 false

3.25 true

3.26 true

3.27 true

3.28 false

3.29 true

SECTION ONE

1.1 Either order:
a. Bible
b. Physical

1.2 intention

1.3 true

1.4 fossils

1.5 theory

1.6 b. few

1.7 b. the whole earth

1.8 a. thorns and thistles

1.9 a. reproduction

1.10 c. a great number

1.11 Any order:
a. grass
b. herbs
c. trees or thistles

1.12 Any order:
a. fish
b. birds
c. cattle
d. reptiles or wildlife

1.13 true

1.14 true

1.15 false

1.16 false

1.17 true

1.18 true

1.19 true

1.20 false

1.21 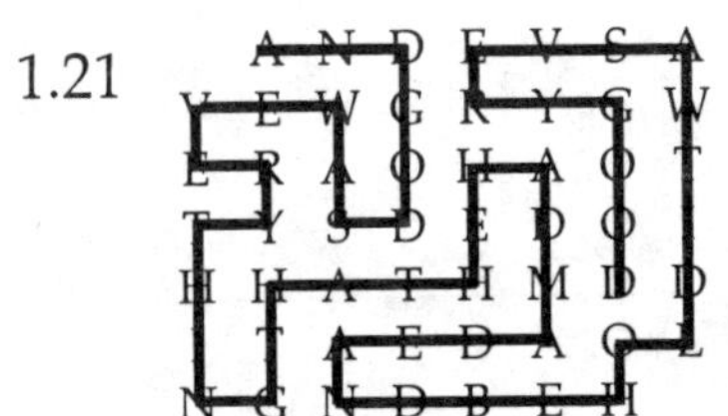

1.22 And God saw everything that He had made and behold, it was very good.

1.23 31

1.24 hardened remains of a plant or an animal

1.25 turned to stone

1.26 Facts; things known about; information.

1.27 offspring; one born of a certain group or family

1.28 made into a fossil

1.29 c

1.30 c

1.31 b

1.32 a

1.33 b

1.34 a

1.35 a. Hint:
There are places on earth with warm climates all year. This could be similar to the climate before the Flood.

b. Hint:
The Arctic areas have times of very cold weather. Mountain regions get very cold, too. Extreme cold was not part of the earth's climate before the Flood.

1.36 a. Hint:
According to fossil forms, there is not as large a variety of animal species today as there was before the Flood.

b. Hint:
Some animal species that lived before the Flood are no longer living.

1.37 a. Hint:
Many of the plants that grew before the Flood were similar to those growing today except in size.

b. Hint:
Some plants grew much larger than their descendants grow.

SECTION TWO

2.1 5400

2.2 450

2.3 900

2.4 75

2.5 33,750

2.6 101,250

2.7 2

2.8 120

2.9 8

2.10 2

2.11 Either order:
a. mist
b. dews

2.12 a year

2.13 clean

2.14 seasons

2.15 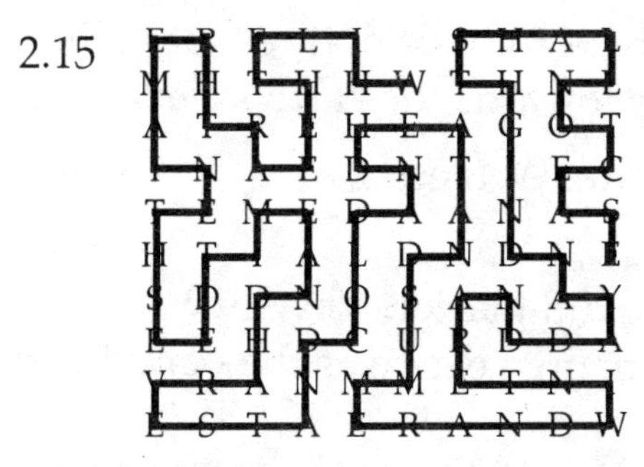

2.16 While the earth remaineth, seedtime and harvest and cold and heat and summer and winter and day and night shall not cease.

2.17 Adult check

2.18 Hints:
a. Was it difficult to build the ark?
b. How did you get the animals to come to the ark?
c. Were you ever frightened during the Flood?
d. How was the earth different after the Flood?

2.19 Adult check

2.20 true

2.21 true

2.22 false

2.23 false

2.24 false

2.25 false

2.26 false

2.27 true

2.28 Adult check – answers will vary

2.29 Hint:
It tells of destruction of animal and plant life.

2.30 Hint:
It did not tell of Noah.

2.31 Hint:
There were no ways for people to write down stories. They were told from parents to children and possibly changed with time.

SECTION THREE

3.1 true

3.2 false

3.3 false

3.4 false

3.5 false

3.6 true

3.7 true

3.8 true

3.9 Examples:
God saw how wicked people had become on the earth. He wanted to make a new start with Noah and his family.

3.10 Examples:
People began to eat animals for meat. Animals became fearful of people. All animals were counted upon to reproduce. Many of them increased their populations greatly.

3.11

D A N E O R A N O
U I D E B L T E F
O T S H A L O K A
L C O C
H E V E
T N T E E M N A N
W I M S A N T B E
O B Y O N E E W T
I D D T H
H T R A E E

3.12 I do set my bow in the cloud and it shall be for a token of a covenant between me and the earth. Genesis 9:13.

3.13 Example:
God said that He would never destroy the whole earth again with a flood. He promised that the seasons would always come in turn. He promised not to curse the world.

3.14 b. complete

3.15 a. nearly all over the world

3.16 b. are extinct

3.17 c. plants were larger once

3.18 crust

3.19 short

3.20 cycle

3.21 moving

3.22 earthquakes

3.23 water

3.24 extinct. No longer living.

3.25 resin. A sticky substance from pine trees.

3.26 crust. Outside covering.

3.27 erode. Wear down with time.

3.28 tradition. Idea or story handed down from parents to children.

3.29 glacier. Large ice mass that constantly moves.

3.30 gravity. A force of nature that pulls things towards the earth's center.

3.31 moisture. Wetness.

3.32 Hints:
Where did glaciers develop?
How many years was the period of the glaciers?
How did glaciers help form the earth's crust?
Will there be a future ice age?
Was there life during the last ice age?
Where did you get your ideas?
Adult check

3.33 g

3.34 d

3.35 a

3.36 h

3.37 c

3.38 e

SECTION ONE

1.1 Adult check

1.2 Hint:
The shapes are similar. The sizes are alike. Special markings (grooves) look similar.

1.3 Hint:
The mold is indented and the bone or shell is opposite. The mold shape is an image of the original item.

1.4 Hint:
The original part leaves a print or copy of itself in the clay.

1.5 Adult check

1.6 Hint:
It has similar shape, size, and markings. Also, it looks like a "twin" to the original.

1.7 Hint:
It is not made from the same substance. The markings are not as clear.

1.8 Hint:
The mold left its print or copy on the new substance. The new substance took the print or shape of the mold.

1.9 Either order:
a. mold
b. cast

1.10 original-remains

1.11 amber

1.12 mummification

1.13 Either order:
a. oil
b. coal

1.14 hard parts (or shell or bones)

1.15 Hint:
Skin and muscle were not decayed. Eyes and other organs were whole. Stomach held undigested food from plants not now in Siberia. It was in a standing position. It was frozen.

1.16 Hint:
List any items omitted from your list in 1.15.

1.17 false

1.18 true

1.19 true

1.20 true

1.21 true

1.22 a

1.23 b

1.24 b

1.25 a

1.26 Across
2. cast
3. remains
6. fossil
7. petrified
8. amber

Down
1. print
2. carbonized
4. mold
5. mineral

1.27 near Agate Springs, Nebraska

1.28 Gobi Desert, Mongolia

1.29 Either order:
a. Arizona
b. Yellowstone

1.30 Sicily

1.31 Maryland

1.32 Los Angeles, California

1.33 Either order:
a. Siberia
b. Alaska

1.34 Baltic Sea area

1.35 Green River, Wyoming

1.36 Alaska

1.37 Massachusetts

1.38 Florissant, Colorado

1.39 Massachusetts

1.40 a. Nebraska
b. Sicily
c. California
d. Siberia
e. Wyoming
f. Mongolia
g. Maryland
h. Alaska
i. Baltic Sea
j. Colorado

1.41 a. Arizona
b. Alaska
c. Yellowstone

1.42 Adult check

1.43 Adult check

1.44 Adult check

1.45 Adult check

1.46 a

1.47 c

1.48 a

1.49 c

1.50 b

1.51 Adult check

1.52 Adult check

SECTION TWO

2.1 true

2.2 true

2.3 true

2.4 false

2.5 true

2.6-2.9 Either order:

2.6 a. flowering plants
b. cone-bearing plants

2.7 a. ferns
b. fungi

2.8 a. red algae
b. green algae

2.9 a. amoeba
b. paramecium

2.10 Any order:
a. insects
b. worms
c. mollusks

2.11 Any order:
a. fish
b. reptiles
c. amphibians
d. birds

2.12 Either order:
a. mammals
b. some fish and reptiles

2.13 seed

2.14 one-celled

2.15 soft

2.16 leaves

2.17 more

2.18 flowers

2.19 false

2.20 false

2.21 true

2.22 false

2.23 true

2.24 false

2.25 true

2.26 false

2.27 Adult check

2.28 c

2.29 b

2.30 b

2.31 b

2.32 a

2.33 b

2.34 b

2.35 Adult check

2.36 b

2.37 c

2.38 c

2.39 a

2.40 c

2.41 a

2.42 Hint:
a. It is a print fossil.
b. It is not large.
c. It looks somewhat like ferns of today.
d. It was probably found at a rock quarry, or a mountain, or any area of sedimentary deposits.
e. It was found in sedimentary rock.

2.43 Hint:
a. The fossil is a fern fossil.
b. Ferns were not large.
c. Ferns lived near water.
d. Ferns needed a climate similar to today's ferns.
e. The fern sank in the sediment and was drowned.

2.44 Adult check

2.45 Adult check

2.46 Adult check

2.47 reconstruct

2.48 man-made

2.49 models

2.50 conclusions, inferences

2.51 scars

2.52 Adult check

SECTION ONE

1.1 true

1.2 true

1.3 false

1.4 true

1.5 false

1.6 false

1.7 true

1.8 Example:
roundness, color, has bump on one side

1.9 Example:
Far-surface looked rather smooth, color was much the same.
Near-surface was rough, patches of color, spots on surface, tiny indentation.

1.10 Distance causes the observer to see the whole object. Surface irregularities are not important. Closeness allows observer to see details.

1.11 We are usually very close to the landforms. (We are smaller.) They appear very large — just as the surface features stood out in the orange at close range. The details of both the orange and the earth's surface are not as obvious at a distance. They are tiny compared to the whole object.

1.12 Rocky Mountains

1.13 Grand Canyon

1.14 Great Plains

1.15 Pacific Ocean

1.16 Mississippi River

1.17 Gulf of Mexico

1.18 Hawaiian Islands

1.19 Appalachian Mountains

1.20 He said that people who trust God will not be moved as Mount Zion is not moved and that God is like mountains around Jerusalem. He is around His people.

1.21 It had hills and valleys.

1.22 He compared the Christian to a city on a hill.

1.23 Example:
1 Corinthians 13:2
The magnitude of faith is compared to the ability to move mountains.

1.24 c. kilometers

1.25 b. mass

1.26 a. center

1.27 d. landform

1.28 a. life

1.29 thinner

1.30 magma

1.31 pressure

1.32 moves

1.33 Example:
The old man's mantle was draped over his shoulders and hung to the ground.

1.34 Example:
Alice was under a great deal of pressure during the difficult test.

1.35 Example:
Allen did not understand the gravity of his actions.

1.36 Example:
Why did you throw away the bread crust?

1.37 Example:
The rocket sped off into outer space.

1.38 Example:
The whole family went to their church to celebrate Mass.

1.39 a
1.40 c
1.41 a
1.42 b
1.43 c
1.44 c
1.45 b
1.46 a
1.47 Adult check
1.48 no — it is or was alive.
1.49 mineral
1.50 mineral
1.51 no — it is synthetic or man-made
1.52 mineral
1.53 no — it is or was alive
1.54 no — sand from different places on earth have different chemical makeups.
1.55 mineral
1.56 mineral
1.57 no — it is or was alive
1.58 a. crystal
1.59 c. flat
1.60 a. six
1.61 a. found in nature
1.62 sandstone
shale
conglomerate
1.63 a. stone
b. pebble
c. silt
d. sand
e. boulder
f. gravel
g. cobble
1.64 c
1.65 a
1.66 a
1.67 c
1.68 b
1.69 b
1.70 a
1.71 c
1.72 a
1.73 a
1.74 a
1.75 b
1.76 b
1.77 a
1.78 Any order:
a. color
b. luster
c. streak
d. hardness
e. cleavage
1.79 Answer should include the five physical tests
1.80 Answer should include the five physical tests.
1.81 Answer should include the five physical tests.
1.82 Adult check

SECTION TWO

2.1 c. weather

2.2 a. sand against

2.3 a. surface

2.4 b. contracting

2.5 d. ice

2.6 c. slow

2.7 gravity

2.8 erosion

2.9 water

2.10 Either order:
a. high mountains
b. cold areas of the earth

2.11 storms

2.12 glaciers

2.13 The water will carry the soil, sand, and pebbles with it to the bottom of the pan.

2.14 Adult check

2.15 The larger ones stayed higher on the board.

2.16 Some of it stayed on the board, but most of it washed into a pile at the end of the board.

2.17 It washed off the board into a pile. Some was still mixed with sand.

2.18 Adult check

2.19 Smaller particles wash away easier.

2.20 false

2.21 true

2.22 false

2.23 true

2.24 true

2.25 true

2.26 true

2.27 Adult check

2.28 Adult check

2.29 b. moving

2.30 a. a fault

2.31 a. bend

2.32 b. moving magma

2.33 a. sedimentary

2.34 Adult check

2.35 Example:
The new company signed a contract to build the long bridge.

2.36 Example:
The teacher introduced another channel of thought.

2.37 Example:
It was not my fault that the cat was locked outside.

2.38 Example:
The crack of the rifle shot startled the young pup.

2.39 Example:
Did you have enough change with you for the parking meter?

2.40 true

2.41 false

2.42 true

2.43 true

2.44 true

2.45 in prison

2.46 Prison foundations shook. All doors were opened. Everyone's bands were loosened.

2.47 The prison keeper was going to kill himself. He trembled. Everyone stayed in prison.

2.48 They told the prison keeper how to be saved. They baptized him. They also talked with his "house" about God.

2.49 Adult check

2.50 c. lava
2.51 c. steam
2.52 a. explosions
2.53 d. squeeze
2.54 b. hot or cold
2.55 d
2.56 f
2.57 b
2.58 a
2.59 e
2.60 b
2.61 c
2.62 f
2.63 God with rock
2.64 a gift and a precious stone
2.65 God with rock
2.66 Faces and a rock
2.67 rock as a good place to build a house
2.68 heart and a firm stone or millstone
2.69 false
2.70 false
2.71 true
2.72 false
2.73 true
2.74 glacier
2.75 lava
2.76 explosions
2.77 erosion
2.78 faulting
2.79 expand

SECTION ONE

1.1 Hint:
The marbles and water cannot take up the same space. The marble displaced the water.

1.2 the large marble

1.3 Hint:
The volume of the large marble is greater than the volume of the small marble.

1.4 Any order:
a. volume
b. mass
c. inertia

1.5 a. matter

1.6 b. volume

1.7 c. properties

1.8 b. mass

1.9 a. Inertia

1.10 b. weight

1.11 Hint:
It floated and did not dissolve

1.12 Hint:
It dissolved in the water.

1.13 Hint:
The ability to dissolve in water — or solubility — is the physical property.

1.14 the sugar

1.15 true

1.16 false

1.17 true

1.18 true

1.19 false

1.20 true

1.21 Either order:
a. physical properties
b. chemical properties

1.22 Either order:
a. ability to burn (combustibility)
b. ability to rust

1.23 Examples:
It is a chemical property. It is the ability of a material to burn. When materials burn, they combine with oxygen to produce other chemicals. There is a chemical change.

1.24 Example:
Rust forms when iron in a material combines in moist air to form iron oxide.

1.25 Example:
Because the force of gravity on earth is greater than the force of gravity on the moon.

1.26 Yes.
Example:
Density is the mass in a given volume. Since your mass and volume are the same on earth as on the moon, your density would be the same.

1.27 solid

1.28 gas

1.29 liquid

1.30 liquid

1.31 physical

1.32 cycle

1.33 Hint:
The water level was highest in Figure 3 and lowest in Figure 2.
Figures 1 and 4 looked about the same.

1.34 The volume of water was the same in each container.

1.35 The shape of the water was the same as the container holding it.

1.36 Hint:
The shape of the container made a difference, and the size of the container was important, too.

1.37 Hint:
It shows liquid takes the shape of its container. It does not have its own shape, but it does have its own size. None of the containers were full.

1.38 Any order:
Solid
pencil
cookie
hammer
fingernail
cotton
needle

Liquid
ink
milk
blood
apple juice

Gas
steam
gasoline fumes
oxygen

1.39 false

1.40 true

1.41 false

1.42 false

1.43 true

1.44 Hint:
Part of the wax has become liquid. (The candle may have gotten smaller.)

1.45 Hint:
Water droplets formed on the plate.

1.46 Hint:
Heated air hit the cold plate. Heat was removed from the air. And the water vapor in the air changed to liquid.

1.47 Hint:
The plate became blackened above the flame.

1.48 Hint:
Even if the candle burns down and appears to get smaller, matter is not lost. It changes into another material — the black substance.

1.49 Hint:
wax to liquid — ph, wax burned — ch, vapor to water — ph, plate blackened — ch

1.50 a. changes state

1.51 a. conservation

1.52 c. mass

1.53 a. mass

1.54 c. appears to get

1.55 Hint:
It will likely melt.

1.56 Hint:
The mass will remain the same.

1.57 Hint:
Mass is conserved in a physical change.

1.58 Hint:
The mass stayed the same, but the matter changed form.

1.59 Hint:
Heat was added to the ice (solid form), causing it to change to liquid.

1.60 Hint:
The prediction and the results should be the same.

1.61 Adult check

1.62 d

1.63 e

1.64 a

1.65 f

1.66 g
1.67 Molecules
1.68 Atoms
1.69 compounds
1.70 motion
1.71 faster
1.72 farther apart
1.73 cycle

SECTION TWO

2.1 false
2.2 true
2.3 true
2.4 true
2.5 false
2.6 Hint:
It does not snow in summer and honor and the fool do not go together, either.
2.7 Hint:
When a fig tree's branch is tender and puts forth leaves, summer is near.
2.8 Hint:
In Figures 1 and 3, I looked more directly at the equator. In Figure 2, North America was more directly in line with me. But in Figure 4, South America was more direct.
2.9 Hint:
At certain times in the earth's orbit, the sun shines more directly on some parts of the earth than on other parts.
2.10 the sun
2.11 Hint:
Figure 2 (or whatever Figure when the sun was shining more directly on North America)
2.12 directly
2.13 winter
2.14 ice
2.15 atmosphere
2.16 b. not agreed upon
2.17 c. shooting stars
2.18 c. coma
2.19 a. elements
2.20 c. not a
2.21 c. telescope
2.22 false
2.23 true
2.24 false
2.25 true
2.26 false
2.27 Adult check
2.28 Adult check
2.29 formation
2.30 decaying
2.31 change
2.32 molecules or matter
2.33 food
2.34 Adult check

2.35 a. Hint:
Not much moisture seems to be in the air.
b. Hint:
The point of temperature when vapor turns to liquid. (droplets of water)
c. Liquid droplets when frozen form snow.
d. Little drops of liquid are droplets.

2.36 Hint:
Tiny cold rocks fall from clouds.

2.37 Hint:
Hot water level went down. Cold water didn't change much.

2.38 Hint:
When heat is applied, the liquid releases molecules and the liquid turns to gas.

2.39 Adult check

SECTION THREE

3.1 the hollow of His hand

3.2 He weighed them.

3.3 a. pass away and come
b. rise and set
c. goes south — then north, whirls continually
d. to the sea, but return again

3.4

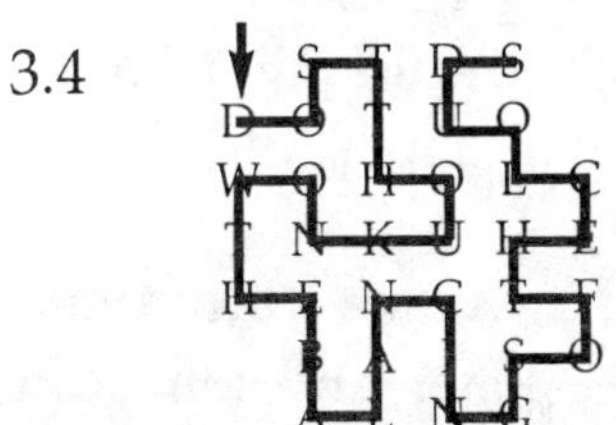

3.5 "Dost thou know the balancings of the clouds. . ."

3.6 ". . .the wondrous works of Him which is perfect in knowledge?"

3.7 Hint:
It tells of the care God took in creation — He balanced the clouds. He has perfect knowledge

3.8 a. water

3.9 c. good judgment

3.10 b. water

3.11 a. controls over it

3.12 b. caused

3.13 a. cattle
b. service of man — food
c. strengthen man's heart
d. refuge for goats

3.14 Lions came out at night — slept by day. Man labored when the sun shone.

3.15 Adult check

3.16 food

3.17 dust

3.18 cycle

3.19 Either order:
a. balance
b. precision

SECTION ONE

1.1 b. unicellular

1.2 b. eukaryote

1.3 c. cell wall

1.4 b. chlorophyll

1.5 a. photosynthesis

1.6 Any order:
   a. animals
   b. plants
   c. fungi
   d. protists
   e. monerans

1.7 cell — the basic unit of all living things.

1.8 cellulose — a substance that forms the walls of plant cells.

1.9 nucleolus — a small part within the nucleus that is very condensed chromatin and consists mainly of RNA and other proteins.

1.10 xylem —The connective tissues in plants that help carry materials through the plant.

1.11 tissue — A group of similar cells connected together that perform similar work.

1.12 fungi — One of the five main kingdoms of living things. They do not produce chlorophyll.

1.13 yeast —A single-celled fungi.

1.14 spores — tiny, specialized structures that are able to grow into a new organism. Spores help an organism survive and move from place to place.

1.15

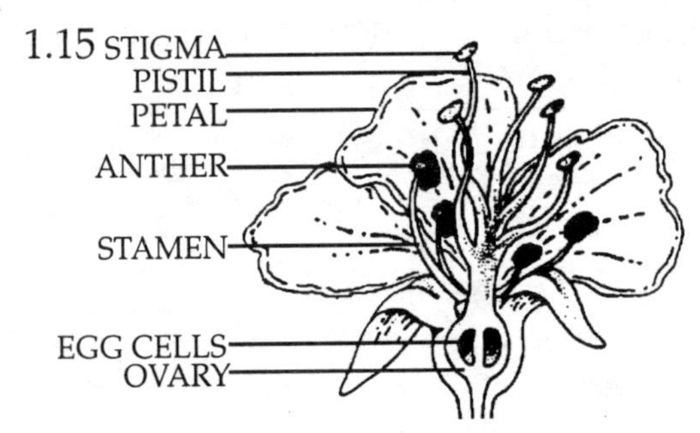

1.16 a

1.17 a

1.18 c

1.19 c

1.20 a

1.21 b

1.22 c

1.23 c

1.24 a

1.25 b

1.26 b

1.27 c

1.28 b

1.29 b

1.30 a

1.31 a. Mitosis is cell division. In mitosis, one cell splits apart to form two new cells. The nucleus brings its materials together to form chromosomes. The chromosomes move in pairs. Soon they pull away from each other, allowing the cell to split into two cells with the same makeup as the original cell.

b. The parent cell develops a bump on itself. Some cytoplasm and some of the nucleus move into the bump. The cell membrane pinches away from the bud. Two cells are produced.

1.32 true

1.33 false

1.34 false

1.35 true

1.36 true

1.37 false

1.38 true

1.39 false

1.40 Adult check

1.41 Adult check

1.42 Adult check

1.43 Mitosis is the process of cell division where a cell divides into two identical cells. Osmosis is the process where materials are brought into cells or out of cells through the cell membrane. They are two unrelated cell processes.

1.44 The host is a plant or animal that supports a parasite. The parasite lives in or on the host and usually feeds off the host's food or the host itself. The parasite needs the host for survival, but the host does not need the parasite.

1.45 They are all stages in a life cycle. They can be identified as stages of development of the animal they will become. The animals in these stages do not have all of the parts that an adult of the species has.

1.46 The nymph appears very much like the adult, but the other two do not. The pupae moves very little, but the others can move a great deal. The larvae sheds its skin as it grows and the others do not shed. The larvae is also wormlike.

1.47 Adult check

1.48 a. a thin layer of hard tissues covering the outside of an animal
b. The dinner plate fell on the floor and broke.

1.49 a. any of the thin, flat plates that protect an animal's body.
b. Example:
The scale on the map helped me decide how far we would travel.

1.50 a. vertebrate

1.51 b. fish

1.52 b. milk

1.53 a. underwater

1.54 c. parents

1.55 b. alligator

1.56 They eat plants – making them primary consumers. They also eat animals – making them secondary consumers.

1.57 Primary consumers usually reproduce more often and have more offspring than secondary consumers. Also, many primary consumers are smaller. Secondary consumers need more primary consumers or they would starve.

1.58 Example:
grass (producer)grasshopper (primary consumer)eagle (secondary consumer)bacteria (decomposer)

1.59 They have plant cells, chlorophyll, and grow in soil.

1.60 They are often small animals. They reproduce at a high rate. They do not have strong jaws or sharp teeth. They are more numerous than secondary consumers.

1.61 They have sharp teeth and claws. Their jaws are very strong. They are less numerous than primary consumers.

1.62 They are microscopic or are fungi. They are not hunters. They feed on dead bodies of organisms.

1.63 true

1.64 true

1.65 true

1.66 true

1.67 true

1.68 true

1.69 true

1.70 c. caring for

1.71 b. sometimes are

1.72 a. photosynthesis

1.73 b. be good stewards of

1.74 a. human

1.75 Adult check

2.1 Noah

2.2 seven

2.3 pressure

2.4 climate

2.5 Either order:
a. oil
b. coal

2.6 Adult check

2.7 b. print

2.8 a. original-remains

2.9 b. the animal is extinct

2.10 a. location

2.11 b. climate of

2.12 It is the process of finding out through reason. Since fossils seem to give data, a decision about early times is made based on the data and reasoning.

2.13 It is a rejoining of fossil parts in such a way that the skeleton of an animal results. Sometimes, a full-sized model of the animal is built.

2.14 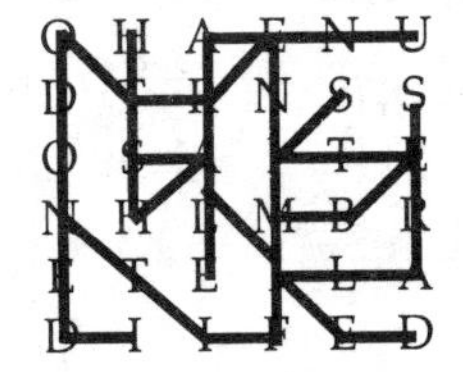

a. amber
b. mineral
c. mastodon
d. unearth
e. shale
f. sites
g. similar
h. identified

2.15 false

2.16 false

2.17 true

2.18 true

2.19 true

2.20 c. geology

2.21 a. crust

2.22 b. melted

2.23 c. metamorphic

2.24 c. hardness

2.25 Adult check

2.26 Adult check

2.27 a. The act of loosening or getting larger.
b. The act of tightening or drawing together.

2.28 When the sun shines on rocks, their surfaces are heated. Heat causes the surfaces to expand slightly. (The inside of the rock does not expand rapidly.) When the surface cools, it contracts. This expanding and contracting can weaken the surface and cause a crack to appear. This is called weathering because temperature is the cause.

2.29 a.-d. Any order (surface forces):
a. weathering
b. erosion
c. glaciers
d. living things

e.-h. Any order (forces below surface):
e. folding
f. faulting
g. earthquakes
h. volcanoes

2.30 Adult check

2.31 Adult check

SECTION THREE

3.1 Energy

3.2 kinetic

3.3 transformation

3.4 lost

3.5 sun

3.6 food

3.7 mechanical

3.8 Burning

3.9 When mechanical energy is being released through an object's movement, the object may move against something else (friction) and some of the mechanical energy is transformed into heat energy.

3.10 It is light energy from the sun.

3.11 Examples:
a. ball hits car
b. photosynthesis
c. current from battery causes flashlight bulb to light
d. dentist drill (running) drills tooth (moves part of tooth)
e. growth of cell causes it to divide
f. heat causes steam — steam moves train — train wheels cause heat on tracks

3.12 true

3.13 false

3.14 true

3.15 true

3.16 true

3.17 false

3.18 true

3.19 false

3.20 Adult check

3.21 e

3.22 g

3.23 f

3.24 a

3.25 i

3.26 b

3.27 j

3.28 c

3.29 k

3.30 l

3.31 Volume is the common property of all matter by which it takes up space. Two objects cannot take up the same space at the same time.

3.32 Hint:
He used just the right amount of matter in water to get it to evaporate at a normal temperature and to condense into clouds for rain.

3.33 The cycles operate only because the properties and structure of matter make it possible.

3.34 life, day and night, wind and water

3.35 a. waters in the hollow of His hand
b. heaven with the span
c. dust in the measure
d. mountains in scales
e. hills in a balance

3.36 c. an orbit

3.37 b. tails

3.38 a. moving

3.39 c. less dense

3.40 c. brightest

3.41 a. predicted

3.42 c. their discoverers

SECTION ONE

1.01 e

1.02 k

1.03 c

1.04 a

1.05 i

1.06 d

1.07 g

1.08 b

1.09 f

1.010 m

1.011 c. cells

1.012 c. discoveries

1.013 b. unicellular

1.014 a. a nucleus

1.015 d. optical microscope

1.016 d. all of these

1.017 b. yolk of an ostrich egg

1.018 The answer should contain some of the following: A cell is the basic unit of all living things. It is the unit of life. All living things that God has created contain cells. A cell contains at least two basic parts: a cell membrane and protoplasm. Three-part cells contain a cell membrane, cytoplasm, and a nucleus.

1.019 The dyes stain certain parts of the cell — such as the cell membrane and the nucleus — so that they stand out more clearly when the cells are viewed under the microscope.

1.020

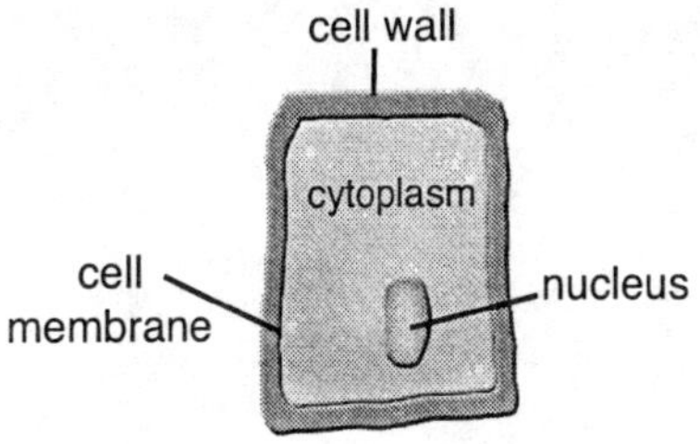

SECTION TWO

2.01 true

2.02 true

2.03 false

2.04 true

2.05 true

2.06 false

2.07 true

2.08 true

2.09 false

2.010 true

2.011 false

2.012 basic unit

2.013 nucleolus

2.014 Bone

2.015 microscope

2.016 leaves

2.017 multicellular

2.018 l

2.019 k

2.020 j

2.021 i

2.022 h

2.023 g

2.024 f

2.025 e

2.026 d

2.027 c

2.028 There are no muscle, nerve and bone cells in plants. Functions of cells are different. Plant cells have a cell wall, chloroplasts, and chlorophyll. Animal cells do not.

2.029 Movement is caused by muscle cells contracting. Small movements happen when only a few muscle cells contract. Large movements result when many muscle cells contract.

2.030 DNA and genes contain the molecular information to make the cells and groups of cells within a living thing what they are to be. The DNA is what makes the offspring of a living thing like the parent.

2.031 Tissues are groups of cells in a multicellular plant or animal that are similar in structure and perform similar functions. The four types of animal tissues are epithelial, muscular, nervous, and connective.

SECTION THREE

3.01 true

3.02 false

3.03 true

3.04 false

3.05 true

3.06 true

3.07 true

3.08 true

3.09 true

3.010 false

3.011 true

3.012 c. discoveries

3.013 d. all of these

3.014 a. organelles

3.015 a. oxygen

3.016 c. White blood

3.017 a. cell division

3.018 Nerve

3.019 Muscle

3.020 Phospholipids

3.021 Any order:
a. DNA
b. RNA
c. other proteins

3.022 Mitosis brings about cell division and two new cells from one original cell. Mitosis starts when the chromatin within the cell begins to rearrange and condense into orderly strands called *chromosomes*. The chromosomes then move into pairs. After that, the chromosome pairs begin to pull apart from each other. Eventually, the chromosome pairs split apart. When they split apart, cell division occurs. Then there are two new cells instead of the original single cell.

3.023 Most of the multicellular plants and animals reproduce themselves by the process known as *male-female reproduction.* A cell from a male parent and a cell from a female parent join together to form a new cell. This process begins a new living thing that has characteristics of both parents. The newly formed cell then begins to reproduce itself through mitosis.

3.024 In photosynthesis, green plants containing chlorophyll absorb energy from the sun to cause a chemical reaction between carbon dioxide and water to produce oxygen and sugars. Respiration is the opposite of photosynthesis. In respiration, oxygen and food combine to produce energy and carbon dioxide and water are given off.

3.025 Microscopes are helpful in viewing cells. There are two basic types of microscopes: optical and electron. Dyes are also helpful in viewing cells so that parts of the cells stand out more clearly.

SECTION ONE

1.01 l
1.02 k
1.03 j
1.04 i
1.05 h
1.06 g
1.07 a
1.08 b
1.09 c
1.010 d
1.011 God
1.012 food
1.013 multicellular
1.014 260,000
1.015 life
1.016 beginning
1.017 growth
1.018 adulthood
1.019 death, or end
1.020 c. chlorophyll
1.021 a. botanists
1.022 c. adulthood
1.023 b. life cycle
1.024 c. both a and b
1.025 Any order:
a. roots
b. leaves
c. stems
d. flowers

SECTION TWO

2.01 b
2.02 c
2.03 a
2.04 b
2.05 b
2.06 b
2.07 a
2.08 c
2.09 c
2.010 a
2.011 true
2.012 true
2.013 true
2.014 true
2.015 true
2.016 false
2.017 true
2.018 true
2.019 false
2.020 false
2.021 beginning
2.022 growth
2.023 adulthood
2.024 death — or end
2.025 b
2.026 c

2.027 b

2.028 b

2.029 a

2.030 flowering

2.031 fertilization

2.032 scales

2.033 seed

2.034 pistil

2.035 Cones do not form fruit. Cones don't have pistils, stamens, anthers, or stigmas.

SECTION THREE

3.01 false

3.02 false

3.03 false

3.04 true

3.05 false

3.06 true

3.07 true

3.08 false

3.09 false

3.010 false

3.011 c

3.012 f

3.013 i

3.014 k

3.015 a

3.016 g

3.017 h

3.018 b

3.019 l

3.020 e

3.021 spore

3.022 prothallus

3.023 fertilization

3.024 adulthood

3.025 in ovaries

3.026 spore cases

3.027 pollen

3.028 by vegetative reproduction

3.029 Example: mold

3.030 A life cycle is the normal stages a plant could be expected to go through if it survived. Not all plants or fungi complete a full life cycle, but a species has a definite pattern of beginning, growth, reproduction, and death.

SECTION FOUR

4.01 false
4.02 false
4.03 false
4.04 false
4.05 true
4.06 false
4.07 true
4.08 true
4.09 false
4.010 false
4.011 false
4.012 true
4.013 true
4.014 true
4.015 false
4.016 c. budding
4.017 c. Algae
4.018 a. Carbon dioxide gas
4.019 b. oxygen
4.020 a. green
4.021 c. cytoplasm
4.022 thread-like bodies of some fungi
4.023 A lump grows on a parent cell. Cytoplasm and part of the nucleus move in. The new cell breaks off.
4.024 Jesus used the example of the lily to tell us not to worry. He compared the splendor of the lily's appearance to that of the splendor of Solomon's clothing.
4.025 A fern's life cycle has an added step. A small, green plant called a prothallus grows from a spore. The prothallus produces sperm and eggs that need to join.
4.026 a.-e. bulbs, stems, underground branching, underground buds/leaves, or spores

SECTION ONE

1.01 d
1.02 a
1.03 b
1.04 c
1.05 c
1.06 e
1.07 b
1.08 a
1.09 c
1.010 c
1.011 false
1.012 false
1.013 true
1.014 true
1.015 true
1.016 true
1.017 false
1.018 false
1.019 false
1.020 true
1.021 b. zoology
1.022 b. herbivores
1.023 c. parameciums
1.024 a. nymph
1.025 c. lives in or on other animals
1.026 c. an invertebrate
1.027 b. a sperm cell
1.028 a. a fly larva
1.029 c. earthworm
1.030 c. insects
1.031 an egg is fertilized
1.032 egg is laid
1.033 nymph is hatched from egg
1.034 wings grow
1.035 adulthood
1.036 An egg cell is fertilized. It is hatched into a larva. Larva changes into adult form. The adult mollusk may reproduce.

SECTION TWO

2.01 true
2.02 false
2.03 false
2.04 true
2.05 true
2.06 true
2.07 false
2.08 true
2.09 true
2.010 false
2.011 e
2.012 a
2.013 b
2.014 c
2.015 c
2.016 d
2.017 c
2.018 a
2.019 b
2.020 e

2.021 c. mammals and birds

2.022 a. Earthworms

2.023 a. testes

2.024 c. reptiles and fish

2.025 b. mammal

2.026 b. gills

2.027 a. only once

2.028 b. bird

2.029 b. many years

2.030 a. inside the mother's body

2.031 a. food, protection and training

2.032 Any order:
a. Only mammals nurse their babies on mother's milk
b. Only mammals have hair.
c. Mammals are warm-blooded.
d. Mammals have a larger, more well-developed brain.
e. Most mammals give their offspring training and protection.

2.033 1. Eggs are laid and fertilized in the water.
2. Tadpoles hatch from eggs. Tadpoles lose tail, grow lungs, grow feet.
3. Frog leaves water to live as a land animal. Adults reproduce.

2.034 a. It protects the yolk and embryo
b. It supplies food for the embryo.
c. It is the part that will grow through mitosis into a chick.

SECTION ONE

1.01 c
1.02 d
1.03 g
1.04 h
1.05 a
1.06 b
1.07 k
1.08 l
1.09 e
1.010 f
1.011 true
1.012 true
1.013 false
1.014 false
1.015 true
1.016 true
1.017 true
1.018 false
1.019 true
1.020 true
1.021 b. evaporates
1.022 c. precipitation
1.023 a. photosynthesis
1.024 a. proteins
1.025 c. birds
1.026 a. leaf
b. caterpillar
c. bird
d. cat
e. bacteria
1.027 The balance of nature is the system where plants and animals depend on each other for some life needs. Plants give off the oxygen required by animals. Animals give off carbon dioxide needed by plants. Plants produce energy. Animals get energy by eating plants or by eating other animals. Organisms need to be eaten to stop over-population of a species.

SECTION TWO

2.01 a
2.02 b
2.03 c
2.04 c
2.05 b
2.06 b
2.07 a
2.08 c
2.09 b
2.010 c
2.011 a. tall grasses
2.012 b. secondary consumers
2.013 b. rainfall
2.014 b. primary consumers
2.015 a. settlers
2.016 a. thorns
2.017 c. fire
2.018 c. tall grass
2.019 c. snake
2.020 c. both plants and animals

2.021 Example:
Sparrows could be sold for money. God was aware when a sparrow died. He cares much more about you. You are worth a great amount to Him.

2.022 A scavenger is an animal that waits for other animals to make a kill. (or) It feeds on dead or rotting organisms.

2.023 Example:
Settlers came and farms were made. New plants and animals were brought in. Most of the bison were killed. Some animals moved to other areas.

2.024 Example:
When a moderate rainfall area had wetter years, many trees began growing. When there were drier years, fires often occurred and killed off the small trees. However, the fires did not kill most grasses. Therefore, prairies with grasslands won out in the open areas. Where the areas were more moist and protected, forests grew.

Rainfall averaged between 25 cm per year and 100 cm per year in the prairie regions. The eastern parts of the prairie usually received more rainfall than the western parts of the prairie. The difference in rainfall is one reason why the food chain varied somewhat from east to west in the prairie ecosystem.

SECTION THREE

3.01 false

3.02 true

3.03 true

3.04 true

3.05 false

3.06 false

3.07 true

3.08 false

3.09 false

3.010 true

3.011 a. producer
b. primary consumer
c. secondary consumer
d. secondary consumer
e. decomposer

3.012 Any five of these six:
a. Walk or take your bicycle instead of taking a car.
b. Place litter in containers.
c. Do not bother animals or their nests.
d. Hunt or fish within the law.
e. Wisely use needed water.
f. Pray that God will show you other ways to be a good steward of His creation.

3.013 b. careful with

3.014 c. precipitation

3.015 c. birds

3.016 a. tall grasses

3.017 c. both harmful and helpful

3.018 a. slowed or stopped

3.019 Example:
We were placed on earth to care for plants and animals in the web of life. We can change it somewhat, but we are not to destroy it.

3.020 When people move into an area, loss of plant and animal life usually occurs. Land is taken over for buildings. Farmers clear land for their crops. As cities begin to develop, large areas of land are covered with concrete. Hunters kill animals for food. They also kill dangerous predators.

People have many uses for the wood of trees. Sometimes, trees stand in the way of roads, farms, or buildings. People remove the trees and this action causes a change for some animals.

SECTION ONE

1.01 true
1.02 true
1.03 false
1.04 true
1.05 true
1.06 true
1.07 true
1.08 true
1.09 false
1.010 false
1.011 false
1.012 true
1.013 true
1.014 true
1.015 false
1.016 d
1.017 e
1.018 a
1.019 b
1.020 f
1.021 c
1.022 b. heat
1.023 b. 50 foot-pounds
1.024 a. heat
1.025 c. kinetic
1.026 c. transformation
1.027 c. having faith in Him
1.028 work
1.029 potential
1.030 kinetic
1.031 God
1.032 solar
1.033 chemical
1.034 Example:
Energy is the ability to do work. There are several energy forms. Energy can be stored as potential energy. Energy that is moving is kinetic energy. Without energy, work would not be done.
1.035 Example:
Work is done when a force moves an object by a distance. It is the movement of matter. Work is done when there is transformation of one form of energy to another.

SECTION TWO

2.01 e
2.02 i
2.03 l
2.04 j
2.05 c
2.06 a
2.07 h
2.08 k
2.09 d
2.010 f

2.011 chemicals
2.012 heat energy
2.013 chemical
2.014 movement
2.015 heat
2.016 friction
2.017 heat
2.018 true
2.019 true
2.020 true
2.021 false
2.022 true
2.023 true
2.024 true
2.025 true
2.026 true
2.027 true
2.028 b. potential
2.029 a. joule
2.030 c. transformation
2.031 b. chemical energy
2.032 Example:
Heated gases cause pressure. The gases are in a chamber with a piston. The piston is moved by the gas pressure. The piston moves other parts of the engine. All of this movement is work.
2.033 Transformation of energy means that it is changed from one form of energy to another. There are many examples that could be given. Some would be chemical to electrical in a battery; mechanical to heat in friction; solar to heat from the sunshine striking an object. There are many more examples that could be given.

SECTION THREE

3.01 true
3.02 false
3.03 true
3.04 true
3.05 false
3.06 true
3.07 true
3.08 true
3.09 true
3.010 false
3.011 true
3.012 true
3.013 false
3.014 true
3.015 false
3.016 a. solar
3.017 b. cost
3.018 b. a distance
3.019 b. Solar cells
3.020 c. fission

3.021 f

3.022 h

3.023 a

3.024 i

3.025 k

3.026 g

3.027 d

3.028 b

3.029 j

3.030 Either order:
a. uses less fuel
b. has much less pollution

3.031 Any of these, any order:
a. Uranium and other radioactive elements are limited and could be used up
b. Harmful radiation; Possibility of nuclear accidents; Hot waste water may damage the environment; Radioactive wastes remain dangerous for long periods of time.

3.032 In fission, the atomic nucleus of an element, such as uranium, is split. The splitting of the nucleus of an atom produces a large amount of energy. Fusion involves the combining of atomic nuclei to form heavier nuclei. In this process, enormous amounts of energy are released.

3.033 Example: Geothermal energy is produced whenever water comes in contact with the hot rocks below the earth's surface. The hot rocks heat the water enough to turn part or all of it into steam. If wells are drilled into the earth to contact this hot water and steam, it can be pumped to the earth's surface. If no underground water and steam exists naturally, water can be pumped down through wells into the ground so that the water is heated by hot rocks. Then the hot water and steam can be used on the surface of the earth to generate electricity or to provide heat energy for other purposes.

SECTION ONE

1.01 d
1.02 e
1.03 m
1.04 h
1.05 i
1.06 j
1.07 c
1.08 b
1.09 n
1.010 g
1.011 false
1.012 true
1.013 false
1.014 false
1.015 false
1.016 true
1.017 false
1.018 false
1.019 true
1.020 false
1.021 Adam and Eve lived in the Garden of Eden.
1.022 Thistles began growing.
1.023 People invented methods of building and instruments for music.
1.024 Most of the people were very wicked.
1.025 God was sorry that He made humans who had become so wicked.
1.026 a
1.027 a
1.028 c
1.029 a
1.030 a
1.031 The people of Noah's time were becoming very wicked. They thought about doing evil all the time. They were not using their abilities for good things. God could use Noah to make a new start.
1.032 Plural of datum; facts; things known about; information
1.033 Offspring; one born of a certain family or group
1.034 Hardened remains of plants or animals.

SECTION TWO

2.01 false
2.02 false
2.03 true
2.04 false
2.05 false
2.06 true
2.07 true
2.08 true
2.09 false
2.010 true

2.011 b
2.012 c
2.013 b
2.014 b
2.015 c
2.016 a
2.017 a
2.018 b
2.019 a
2.020 a
2.021 a
2.022 a
2.023 a
2.024 c
2.025 c
2.026 c
2.027 b
2.028 b
2.029 a
2.030 b
2.031 turned to stone
2.032 exactness, rightness
2.033 Hint:
Large deposits of animal fossils have been found in caves, deep cracks, under soil, and on mountainsides. They were not weathered, burned, or chewed. Destruction had to be quick. Movement of soil and stones can be explained by the Flood, too.
2.034 Hint:
The people need to tell stories instead of writing them. They may have changed with time.

SECTION THREE

3.01 false
3.02 true
3.03 true
3.04 true
3.05 false
3.06 true
3.07 true
3.08 true
3.09 true
3.010 true
3.011 a
3.012 d
3.013 c
3.014 d
3.015 d
3.016 a
3.017 b

3.018 A force of nature that pulls things toward the earth's center.

3.019 Wear down with time.

3.020 Information or facts known about.

3.021 Not living anymore — such as a group of birds that had no offspring, then died.

3.022 b. only after

3.023 b. Bible

3.024 b. eaten by humans

3.025 b. changed earth

3.026 c. permafrost

3.027 a. plants

3.028 c. water cycle

3.029 a. human lives

3.030 The mammoth fossils have been found all over the world. Some of them were so complete that food in their stomachs was not digested. The food and the placement of fossils indicate a worldwide mild climate. Also, there were many fossils showing huge size and suggesting that there was a great abundance of food.

3.031 Glacier theories state that glaciers formed rock and soil layers. The Flood on the other hand is thought to have formed the layers. The glacier theories say that there was an ice age that covered the earth for millions or billions of years. The Flood theory suggests that the ice age started after the Flood and only lasted a short while, starting the water cycle.

3.032 Examples:
Gravity pulled the waters to the lowest parts of the earth. The waters flowed down the surface of the earth as continents divided and ocean basins were deepened. Glaciers were forming. The water cycle took effect.

SECTION ONE

1.01 false
1.02 false
1.03 true
1.04 true
1.05 true
1.06 true
1.07 true
1.08 false
1.09 true
1.010 true
1.011 true
1.012 b
1.013 a
1.014 b
1.015 d
1.016 c
1.017 c
1.018 a
1.019 a
1.020 b
1.021 b
1.022 a
1.023 b
1.024 b
1.025 a
1.026 c
1.027 a
1.028 c
1.029 b
1.030 Any order:
a. print fossils
b. original-remains fossils
c. petrified fossils
d. carbonized fossils

1.031 Either order:
a. coal
b. oil

1.032 Hint:
An animal was trapped in mud and died. The soft parts decayed, leaving the hard parts. The mud hardened around the remains. Later, minerals dissolved the original material, leaving a mold of the original remains. Minerals could also come into the mold, harden, and become a cast.

1.033 Hint:
They can take care not to damage fossils they find. Also, they can take care of places where fossils are located.

SECTION TWO

2.01 e
2.02 f
2.03 g
2.04 a
2.05 b
2.06 c
2.07 d
2.08 h
2.09 k
2.010 l

2.011 false
2.012 true
2.013 false
2.014 true
2.015 true
2.016 false
2.017 true
2.018 true
2.019 false
2.020 false
2.021 false
2.022 b
2.023 c
2.024 c
2.025 a
2.026 b
2.027 b
2.028 Any order:
a. print
b. original-remains
c. petrified
d. carbonized
2.029 Many of them would say that the dating methods used to date ancient objects are not accurate or reliable. Creation scientists also point out that some "younger" fossils have been found in soil or rock layers below the "older" fossils.
2.030 Hint:
By comparing fossils with living things and by weighing each bit of new information

1.01 d

1.02 e

1.03 i

1.04 a

1.05 b

1.06 c

1.07 f

1.08 true

1.09 false

1.010 false

1.011 true

1.012 true

1.013 true

1.014 false

1.015

crust
inner core
outer core
mantle

1.016 c. shale

1.017 b. the Flood

1.018 b. limestone

1.019 c. color

1.020 a. lava

1.021 b. constantly changing

1.022 b. Crystals

1.023 Any order:
a. mountains, valleys, plains
b. hills, oceans, rivers
c. lakes, plateaus

1.024 Any order:
a. color
b. streak
c. cleavage
d. luster
e. hardness

1.025 Heat and pressure can force magma towards the earth's surface. When magma gets near enough to the surface to cool down, igneous rocks form from the cool magma.

1.026 Heat and pressure from within the earth cause igneous and sedimentary rocks to change physically (in looks) and chemically (new material).

1.027 Sediment that builds up on the earth is cemented together by the pressure of water or other sediments.

SECTION TWO

2.01 f

2.02 e

2.03 g

2.04 a

2.05 b

2.06 j

2.07 k

2.08 l

2.09 c

2.010 d

2.011 true

2.012 false

2.013 true

2.014 true

2.015 false

2.016 false

2.017 true

2.018 true

2.019 false

2.020 true

2.021 c. living conditions

2.022 b. cutting lumber

2.023 c. weathering

2.024 b. expand

2.025 d. deserts

2.026 a. delta

2.027

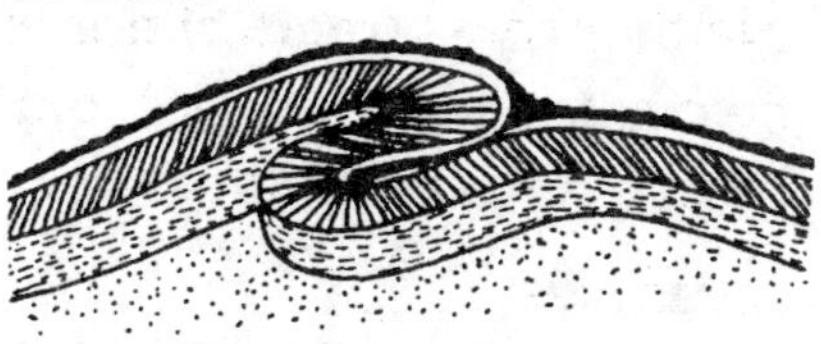

2.028

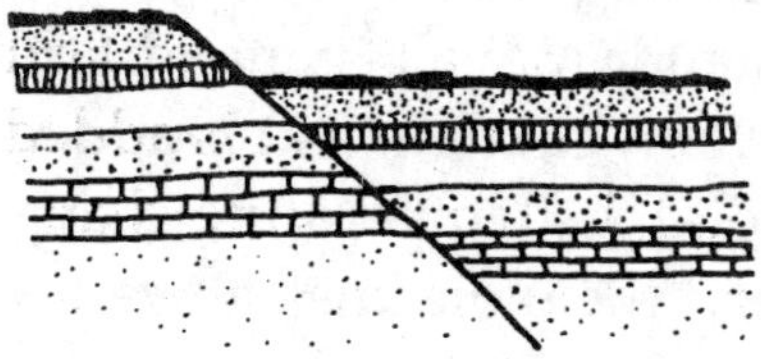

2.029 flattened

2.030 igneous

2.031 magma

2.032 nickel

2.033 They are formed as a result of magma getting near the surface and cooling. They can come from volcanoes, or through magma forcing itself between rock layers.

2.034 A mineral describes a substance that has four features. (1) A mineral is found in nature. Synthetic or man-made substances are not minerals. (2) A mineral has the same chemical makeup wherever it is found on the earth. (3) The atoms of a mineral are arranged in a regular pattern and form solid units called *crystals.* (4) Almost all minerals are made up of substances that were never alive.

SECTION ONE

1.01 b

1.02 a

1.03 b

1.04 a

1.05 b

1.06 a

1.07 b

1.08 c

1.09 b

1.010 c

1.011 true

1.012 false

1.013 true

1.014 false

1.015 false

1.016 true

1.017 true

1.018 false

1.019 false

1.020 a. breaking it

1.021 c. inertia

1.022 b. physical

1.023 a. H_2O

1.024 a. element

1.025 c. cycle

1.026-1.028 Any order:
1.026 solid
1.027 liquid
1.028 gas

1.029 solid

1.030 physical

1.031 chemical

1.032 the same

1.033 Atoms

1.034 Neither mass (matter) nor energy can be created or destroyed, but each may be converted into the other.

1.035 Examples:
Adding heat makes the molecules of matter move faster and farther apart. When enough heat is added to a solid, its temperature changes enough and the molecules move fast enough for the solid to change to a liquid. Adding more heat will eventually change the temperature more and make the molecules move even faster so that the liquid is changed to a gas.

SECTION TWO

2.01 false

2.02 true

2.03 true

2.04 false

2.05 true

2.06 false

2.07 false

2.08 true

2.09 b

2.010 l

2.011 e

2.012 i

2.013 a

2.014 c

2.015 k

2.016 h

2.017 m

2.018 f

2.019 j

2.020 g

2.021 a. predict

2.022 c. axis

2.023 a. brighter

2.024 c. dew point

2.025 a. water vapor is

2.026 a. forms

2.027 b. chemical

2.028 a. physical

2.029 a. solid
b. liquid
c. gas

2.030 Hint:
Seasons are cycled through nature. They can be predicted.

2.031 Hint:
The comets orbit the sun. They can be predicted.

SECTION THREE

3.01 e

3.02 a

3.03 j

3.04 k

3.05 d

3.06 c

3.07 b

3.08 l

3.09 g

3.010 h

3.011 false

3.012 true

3.013 false

3.014 false

3.015 true

3.016 true

3.017 true

3.018 false

3.019 false

3.020 false

3.021 a. tail

3.022 b. an orbit

3.023 c. not certain

3.024 b. direct

3.025 a. rises

3.026 b. change

3.027 c. gas

3.028 c. a compound

3.029 shorter

3.030 brighter

3.031 Hint:
Physical changes in water cause the water cycle to function. The liquid changes to gas form and rises with warm air. After it cools, clouds of liquid are formed and it drops back to earth.

3.032 Hint:
Minerals and other matter are part of something living, but are deposited back to earth through decay. Then these minerals and matter can be used again by living things.

3.033 Hint:
Mass (matter) is neither created nor destroyed. Changes may take place, but they neither add nor take away matter.

3.034 Example:
Verse 4: "One generation passeth away and another cometh" shows life, reproduction, death cycle.
Verse 5: rising and setting of sun — daily cycle.
Verse 6: wind cycles explains general weather patterns and wind movements.
Verse 7: water cycle through rivers to sea — evaporation — rain — rivers

SECTION ONE

1.01 a
1.02 g
1.03 e
1.04 c
1.05 a
1.06 b
1.07 f
1.08 c
1.09 d
1.010 c
1.011 a. five kingdoms
1.012 b. multicellular
1.013 b. photosynthesis
1.014 c. hair
1.015 producer, primary consumer, secondary consumer, decomposer
1.016 true
1.017 false
1.018 false
1.019 false
1.020 false
1.021 true
1.022 true
1.023 true
1.024 false
1.025 true
1.026 The balance of nature depends on the carbon cycle. Animals need the oxygen given off by plants and plants need the carbon dioxide given off by animals.
1.027 Most plants and animals reproduce cells through mitosis — cell division. Without it, growth could not happen.

SECTION TWO

2.01 false
2.02 true
2.03 true
2.04 true
2.05 true
2.06 true
2.07 false
2.08 true
2.09 false
2.010 false
2.011 a. crush stone
2.012 a. fault
2.013 c. lava
2.014 c. layers
2.015 b. weathering
2.016 j
2.017 g
2.018 h
2.019 m
2.020 l
2.021 d
2.022 b
2.023 c
2.024 f

2.025 n

2.026 Any order:
a. crust
b. mantle
c. core (inner, outer)

2.027 As the Flood covered the earth, the bodies of animals could have been washed into low places and covered with sediment.

2.028 As the Flood waters moved over the earth, soil and rocks could have moved with the water. Also, pressure from the water could have caused earthquakes.

SECTION THREE

3.01 b. potential

3.02 c. mechanical

3.03 a. work

3.04 b. good stewards

3.05 e

3.06 d

3.07 i

3.08 a

3.09 g

3.010 h

3.011 b

3.012 false

3.013 true

3.014 false

3.015 true

3.016 true

3.017 true

3.018 false

3.019 true

3.020 false

3.021 false

3.022 true

3.023 true

3.024 true

3.025 true

3.026 Any order:
a. solid
b. liquid
c. gas

3.027 Any four of the following, any order:
a. color
b. odor
c. density
d. brittleness, or conductivity, or solubility

3.028 Inertia is a common property of all matter. Inertia means that an object at rest will stay at rest, or in motion will stay in motion, unless acted upon by an outside force.

3.029 Neither mass (matter) nor energy can be created or destroyed, but each may be converted into the other.

Science 501
LIFEPAC Test

1. d
2. e
3. f
4. b
5. c
6. g
7. h
8. i
9. j
10. k
11. true
12. true
13. true
14. true
15. false
16. true
17. false
18. false
19. true
20. true
21. c. cells
22. a. a nucleus
23. c. oxygen and sugars
24. d. red blood cells
25. b. nuclear membrane
26. Any order:
    a. optical
    b. electron
27. Any order:
    a. nuclear membrane
    b. chromatin
    c. nucleolus
28. Any order and any four of the following six:
    white blood cells
    red blood cells
    nerve cells
    bone cells
    epithelial cells
    muscle cells
29. The answer should contain some of the following: A cell is the basic unit of all living things. It is the unit of life. All living things that God has created contain cells. A cell contains at least two basic parts: a cell membrane and protoplasm. Three-part cells contain a cell membrane, cytoplasm, and a nucleus.
30. Food is brought into the body through eating and the body's digestive system. Oxygen is brought into the body through breathing. Respiration occurs when the food is combined with the oxygen in the body, giving off energy the body needs to perform life and work.

Science 502
LIFEPAC Test

1. b
2. a
3. b
4. c
5. c
6. b
7. b
8. c
9. c
10. b
11. a
12. b
13. c
14. b
15. b
16. vegetative
17. fruit
18. stalk
19. c. fertilization
20. a. mitosis
21. c. pollen
22. b. cutting and seed
23. c. adult stage
24. flower matures
25. bees help
26. pollen makes tube in style
27. fertilization
28. petals dry up
29. fruit is formed
30. seed moves
31. To protect seeds. To provide for seeds. To provide food for animals. To help seeds move.
32. Through budding. A lump forms. Cytoplasm moves into the new cell. The cell breaks away.

Science 503
LIFEPAC Test

1. g
2. h
3. i
4. l
5. j
6. a
7. b
8. c
9. d
10. e
11. five
12. move around
13. cycle
14. embryo
15. parasite
16. b. lizards and snakes
17. c. 2,000 miles
18. a. a fly larva
19. c. testes
20. b. paramecium
21. egg is fertilized
22. egg hatches
23. larva grows
24. pupa forms
25. becomes adult
26. The hosts for the tapeworm larvae are usually fleas. The host fleas are infested with the tapeworm larvae. When a cat cleans itself, it swallows the fleas. The tapeworm larvae on the fleas change into tiny tapeworms. They then live and grow as parasites in the intestines of the cat. The tapeworms produce eggs and fertilize them with sperm. The fertilized eggs are carried out of the cat's body as waste.

    Fleas feed on animal waste. If tapeworm eggs are in the waste, the eggs are brought into the fleas where they hatch into larvae. When the cat swallows the larvae-infested fleas, the tapeworm life cycle continues.
27. true
28. true
29. false
30. true
31. true
32. false
33. false
34. true
35. false
36. true

Science 504
LIFEPAC Test

1. a
2. d
3. c
4. a
5. b
6. d
7. c
8. b
9. d
10. a
11. true
12. true
13. false
14. true
15. false
16. false
17. true
18. false
19. true
20. true
21. c. roots
22. b. ecosystem
23. c. stewardship
24. c. pollution
25. a. terrarium
26. a. bison
27. Adult check
28. A balance of nature occurs when the life needs of all the living things in an area of the earth are met.
29. According to Jesus, you are much more important to God than a whole flock of sparrows.

Science 505
LIFEPAC Test

1. g
2. h
3. i
4. j
5. k
6. a
7. b
8. c
9. l
10. d
11. true
12. true
13. true
14. true
15. false
16. true
17. true
18. true
19. false
20. false
21. b. chemical
22. a. 50 foot-pounds
23. c. transformation
24. c. having faith in Him
25. a. pollution
26. c. nuclear fusion
27. c. hot rocks
28. b. political power
29. energy
30. force
31. heat
32. steward
33. Transformation of energy means that it is changed from one form of energy to another. There are many examples that could be given. Some would be chemical to electrical in a battery; mechanical to heat in friction; solar to heat from the sunshine striking an object. There are many more examples that could be given.
34. The sun's light rays are free. Solar energy is a clean energy source that does not produce pollution.

Science 506
LIFEPAC Test

1. c
2. c
3. b
4. b
5. c
6. b
7. c
8. a
9. b
10. c
11. b
12. c
13. true
14. true
15. true
16. false
17. false
18. false
19. false
20. true
21. true
22. false
23. true
24. Any order:
    a. glaciers
    b. water pressure forced land to move
    c. gravity pulled water to the lowest places, forcing mountains to form; or wind, rain, shifting land masses
25. Any order:
    a. continent drift
    b. earthquakes
    c. volcanoes
    or wind, rain
26. a
27. b
28. c
29. b
30. b
31. a

Science 507
LIFEPAC Test

1. true
2. false
3. true
4. false
5. false
6. true
7. false
8. true
9. false
10. true
11. true
12. true
13. d
14. e
15. f
16. i
17. j
18. a
19. b
20. c
21. k
22. l
23. a
24. b
25. c
26. c
27. c
28. b
29. Any order:
    a. fossil type
    b. plant, animal, or living thing type
    c. age
30. Either order:
    a. mold
    b. cast
31. It is the remains of an ancient plant, animal, or other living thing that was once alive.
32. Example:
    Water containing minerals soaked into the original wood. If some or all of the original wood remains, it is called permineralization. If the minerals totally replace the original wood, it is called replacement.

Science 508
LIFEPAC Test

1. false
2. true
3. true
4. false
5. true
6. true
7. false
8. true
9. false
10. true
11. f
12. h
13. g
14. i
15. k
16. j
17. b
18. e
19. c
20. d
21. m
22. a. outer core
    b. crust
    c. mantle
    d. inner core
23. a. 90 minutes
24. c. shale
25. a. the Flood
26. b. California
27. c. living conditions
28. Any order:
    a. color
    b. streak
    c. cleavage
    d. luster
    e. hardness
29. Earthquakes usually cause wide cracks in the earth's surface. Landslides may also result from earthquakes.
30. Erosion sometimes carries good soil away. Without good soil, plants do not grow well. Crops are poorer. Less food is produced.

Science 509
LIFEPAC Test

1. false
2. true
3. true
4. true
5. false
6. false
7. true
8. false
9. true
10. m
11. f
12. a
13. e
14. h
15. b
16. i
17. c
18. l
19. k
20. d
21. j
22. Any order:
    a. volume
    b. mass
    c. inertia
23. b. density
24. b. heat
25. c. new
26. c. shape
27. b. cycled
28. c. inertia
29. b. CO_2
30. c. Ecclesiastes, chapter 1
31. Neither mass (matter) nor energy can be created or destroyed, but each may be converted into the other.
32. Hint:
    He used just the right amount of matter in water to get it to evaporate at a normal temperature and to condense into clouds for rain.

Science 510
LIFEPAC Test

1. e
2. f
3. a
4. g
5. b
6. h
7. j
8. c
9. l
10. k
11. false
12. false
13. true
14. true
15. true
16. true
17. true
18. false
19. false
20. true
21. false
22. false
23. 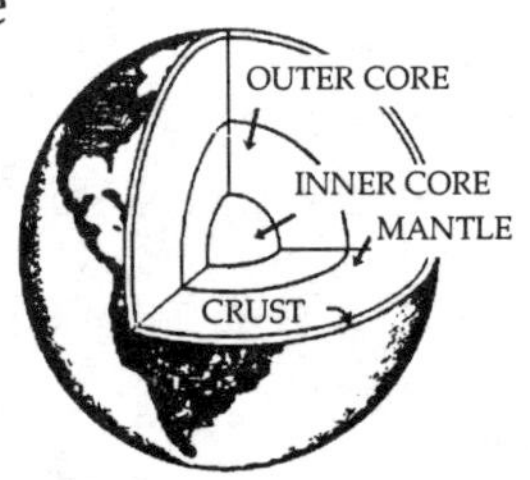

24. a. invertebrate
25. b. secondary consumer
26. a. five kingdoms
27. c. carbonized
28. b. fault
29. c. God's order
30. Any two of the following:
    (1) Only mammals nurse their babies on the mother's milk and (2) only mammals have hair. (3) Baby mammals are very weak at birth and need the care, protection, and training of their parents. (4) The brains of mammals are larger than other animals. (5) Mammals, like birds, are warm-blooded.
31. Some of the following examples should be mentioned:
    Human beings can affect the balance of nature. God has given us the ability to make decisions. We can decide to preserve nature or destroy it. These decisions can be very important for the balance of nature. Loss of life and pollution can be the results of wrong decisions.
    The decisions to build homes or clear land can affect the balance of nature. Some species can no longer survive after land is cleared or homes are built. Other plants and animals may increase in numbers with such human-caused changes. In such circumstances, human beings may need to become *predators* in order to keep the balance of nature.
    Plant and animal life are affected by pollution. Waste chemicals in the air can slow growth in plants and cause problems with photosynthesis. Lungs of animals can be harmed and lives shortened. Chemicals polluting the water can cause similar problems for plants and often poison animals. These human influences work against the natural balance of nature.
    God has given human beings responsibility for life (Genesis 1:28). We have been told to care for other living things. God wants us to be good *stewards* of His creation. Our choices determine whether we will have good stewardship of these things.

1. true
2. false
3. true
4. true
5. true
6. true
7. false
8. false
9. false
10. true
11. Any order:
    a. carbon dioxide
    b. water
    c. sunlight
12. a. - b. Either order:
    a. oxygen
    b. food
    c. carbon dioxide
13. a. oxygen
    b. carbon dioxide
14. a. fearfully
    b. wonderfully
15. need drawing and label
16. c. microscope
17. b. male-female reproduction
18. a. organelles
19. a. cell division
20. b. DNA
21. Examples; any order:
    a. epithelial
    b. connective
    c. muscle or nervous
22. Either order:
    a. oxygen
    b. sugars or food
23. e
24. f
25. a
26. g
27. b
28. h
29. c
30. i
31. d
32. l

1. false
2. false
3. true
4. true
5. true
6. true
7. true
8. false
9. true
10. false
11. true
12. true
13. true
14. false
15. false
16. a. spore begins growth
    b. prothallus forms
    c. sperm and egg develop
    d. fertilization
    e. tiny fern grows
    f. fern matures
    g. many spores are formed
17. mitosis
18. pollen
19. fertilization
20. growth
21. pollen
22. Example:
    Some seed plants and ferns can reproduce when a part of the plant forms roots and grows a new plant. Parts that can grow are roots, stems, leaves, bulbs, and underground branching.
23. Example:
    They are carried by wind, animals, water, bees, birds and humans.
24. a. - c. Any order:
    a. ovary
    b. pistil
    c. stigma, egg

    d. - e. Either order:
    d. stamen
    e. anther, pollen, sperm
25. i
26. g
27. h
28. f
29. a
30. k
31. b
32. l
33. e
34. c
35. j

1. true
2. true
3. false
4. true
5. false
6. false
7. true
8. false
9. false
10. true
11. parasite
12. amphibians
13. an embryo
14. testes
15. tiny hairs
16. a. fertilized
17. c. tadpoles
18. c. 1 million
19. b. infested fleas
20. c. mammals
21. a. egg cell is fertilized
    b. egg hatches
    c. in tadpole form
    d. tail is lost
    e. adult eats insects
22. Spiders do not have six legs. They have only two body parts. There are no wings.
23. A mollusk is a soft-bodied invertebrate animal that has no bones. Most species of mollusks grow hard shells to protect themselves.
24. Fish breathe by means of gills. They move through water with fins and a tail.
25. An amphibian is an animal that spends part of its life in the water and part of its life on land.
26. Mammals must care for, protect, and train their offspring in order for the offspring to survive.

1. true
2. false
3. false
4. true
5. false
6. false
7. true
8. true
9. false
10. true
11. d
12. a
13. j
14. b
15. e
16. l
17. k
18. f
19. h
20. m
21. fire
22. grazers
23. settlers
24. predators
25. air pollution
26. photosynthesis
27. Example:
    The needs of all life webs are supplied from the carbon, water, and chemical cycles. Food chains have producers, consumers and decomposers.
28. Example:
    God asks that each person take care of His creation of life. He wants us to be careful (to be good stewards).
29. Example:
    Hunting can harm the balance of nature when too many animals are shot or when hunters kill off all predators. Hunting is part of the balance of nature when it is done within the law. Humans are just another predator in this case.
30. b. careful with
31. a. c. oxygen
    b. b. carbon dioxide
32. Any order:
    a. a. air   b. c. water
33. a. grasshoppers
34. a. fungi
35. c. dead animals
36. c. web of life

1. false
2. true
3. false
4. true
5. true
6. true
7. false
8. true
9. true
10. false
11. a. heat
12. a. lost
13. c. pollution
14. c. hot rocks
15. b. potential
16. c. chemical
17. c. work
18. b. heat
19. j
20. h
21. f
22. d
23. b
24. a
25. c
26. e
27. g
28. i
29. Transformation of energy means that it is changed from one form of energy to another. There are many examples that could be given. Some would be chemical to electrical in a battery; mechanical to heat in friction; solar to heat from the sunshine striking an object. There are many more examples that could be given.
30. Example:
    Supplies of some energy sources like coal, oil, and natural gas are running low. We must be careful not to waste them and to use them wisely. We must also use other energy sources if we can, so that there will be enough energy for the future.
31. fusion
32. chemical
33. chemical
34. work
35. Any two of these, any order:
    Uranium and other radioactive elements are limited and could be used up. Harmful radiation; Possibility of nuclear accidents; Hot waste water may damage the environment; Radioactive wastes remain dangerous for long periods of time.
36. Automobiles and electrical power generation.

1. false
2. false
3. true
4. true
5. false
6. true
7. true
8. true
9. true
10. true
11. b
12. a
13. c
14. a
15. a
16. c
17. a
18. b
19. c
20. a. worldwide mild climate
    b. much water covered the earth
    c. present water cycle formed
    d. glaciers formed
    e. seasons affected plant growth
21. ark
22. agrees
23. destruction or loss
24. written
25. water
26. permafrost
27. continent
28. The people of Noah's time were becoming very wicked. They thought about doing evil all the time. They were not using their abilities for good things. God could use Noah to make a new start.
29. Large deposits of animal fossils have been found in caves, deep cracks, under soil, and on mountainsides. They were not weathered, burned, or chewed. Destruction had to be quick. Movement of soil and stones can be explained by the Flood, too.
30. b. fossils
31. a. animals
32. b. volcanoes

1. b
2. l
3. h
4. a
5. f
6. j
7. i
8. k
9. g
10. c
11. false
12. false
13. true
14. true
15. false
16. true
17. true
18. false
19. true
20. true
21. b. Carbonized
22. a. safety
23. c. seed-bearing plants
24. a. conclusions
25. Any order:
    a. fossil type
    b. plant, animal, or living thing type
    c. age
26. Either order:
    a. mold
    b. cast
27. Either order:
    a. limestone
    b. shale or sandstone
28. Example:
    Fossil hunters must take care that fossils are not damaged or destroyed, making it more difficult to identify evidence of ancient life and learn of God's greatness.
29. Example:
    Models tell us more about life in the past than a pile of bones. They give us a better idea of size, shape, and function.
30. It is the remains of an ancient plant, animal, or other living thing that was once alive.
31. Example:
    Fossils have been found all over the world. Sediments held many fossils, but tar pits, amber, permafrost, caves, and coal held fossils, too.

1. b
2. l
3. a
4. k
5. j
6. i
7. f
8. g
9. d
10. c
11. true
12. true
13. true
14. true
15. false
16. false
17. true
18. true
19. true
20. true
21. a. sphere
22. a. smaller
23. b. minerals
24. c. fault
25. a. pressure
26. b. temperature
27. b. shape
28. a. lakes
29. Example:
    Folding can be pushing mountains up while erosion and weathering are wearing down the rocks on the mountains.
30. Example:
    Folding has caused some land areas to be pushed upward. Parts that were once under sea are no longer able to support life.
31. Example:
    Heat and pressure force magma to the earth's surface. As the magma cools, it forms igneous rock (granite and quartz).
32. Example:
    shaking or cracking the ground, rumbling, land moving in different directions, landslides, tidal waves.
33. Any order:
    a. color
    b. luster
    c. hardness
    d. streak or cleavage

1. c
2. c
3. b
4. a
5. c
6. a
7. c
8. b
9. b
10. a
11. true
12. false
13. false
14. true
15. true
16. false
17. false
18. true
19. true
20. Examples:
    Any order:
    a. coma
    b. nucleus
    c. tail
    d. hydrogen cloud
21. c inertia
22. b. volume
23. b. slanting
24. b. created or lost
25. a. water
26. b. dust
27. c. the dew point
28. Example:
    Both physical and chemical changes may take place. Chemically, the matter changes to another material. Physically, the matter may change to liquid or gas form.
29. Example:
    It shows that God was precise in Creation. He cared about what He was doing. It also shows His perfect knowledge.
30. Example:
    The earth is tilted on its axis. As it moves around the sun, the tilt causes part of the earth to receive the sun's rays more directly (summer). For the part of the earth tilted away from the sun, it is winter.
31. Example:
    Adding acid or base to some matter can cause it to change chemically. Chemical change produces new material. Rusting and burning are examples. Physical change is changing matter from one form to another.
32. Example:
    Living things are formed, grow, die, and decay. Bacteria is in the soil and in the decaying process. Some matter returns to the soil where it can be used as food for new plants.

1. false
2. true
3. true
4. false
5. true
6. false
7. true
8. true
9. true
10. true
11. h
12. d
13. j
14. g
15. a
16. k
17. e
18. i
19. l
20. b
21. c. one-celled
22. c. shape
23. a. folding
24. b. left out of
25. a. chemically
26. c. solubility
27. Any order:
    a. chemical
    b. sound
    c. mechanical
    d. light
    e. electrical or heat
28. Any order:
    a. plains
    b. valley
    c. mountains
    d. hills or lakes or rivers
29. Example:
    When flood waters washed soil and rocks, layers could have settled. Then later, other layers could have been washed in.
30. Example:
    It is the changing of one energy form to another. Work happens during the transformation.
31. Mitosis is important for growth